The Future of Europe

Series Editors

Michael Kaeding , Institut für Politikwissenschaft, Universität Duisburg-Essen, Duisburg, Nordrhein-Westfalen, Germany

Senem Aydin-Düzgit, Faculty of Arts and Social Sciences, Sabanci University, Istanbul, Turkey

Johannes Pollak, Webster Vienna Private University, Vienna, Wien, Austria

The "Future of Europe" series consists of monographs and edited volumes analyzing topical European issues from the perspective of each EU Member State and neighboring countries, helping to understand the different aspects of the future of the European project. It aims at combining two goals: high quality research-based and/or informed contributions stimulating pan-European national and European, academic and non-academic discussions around the "Future of Europe", involving preferably leading academic scholars and practitioners.

The series provides an authoritative library on the Future of the European Union ranging, amongst others, from general conceptual texts to assessments of countries, regions, key institutions and actors, policies and policy processes. Books in the series represent up-to-date sought-after sources of information and analysis reflecting the most up-to-date research and assessments of aspects related to the Future of Europe. Particular attention is paid to accessibility and clear presentation for a wide audience of students, practitioners and interested general readers.

Michael Kaeding • Johannes Pollak •
Paul Schmidt

Editors

Climate Change and the Future of Europe

Views from the Capitals

 Springer

Editors
Michael Kaeding (iD)
University of Duisburg-Essen
Duisburg, Germany

Johannes Pollak
Webster Vienna Private University
Vienna, Wien, Austria

Paul Schmidt
Institute for Advanced Studies and
Austrian Society for European Politics
Vienna, Austria

ISSN 2731-3379   ISSN 2731-3387 (electronic)
The Future of Europe
ISBN 978-3-031-23327-2   ISBN 978-3-031-23328-9 (eBook)
https://doi.org/10.1007/978-3-031-23328-9

This Springer imprint is published by the registered company Springer Nature Switzerland AG
The registered company address is: Gewerbestrasse 11, 6330 Cham, Switzerland

TEPSA
Trans European Policy Studies Association

Co-funded by
the European Union

The European Commission's support for the production of this publication does not constitute an endorsement of the contents, which reflect the views only of the authors, and the Commission cannot be held responsible for any use which may be made of the information contained herein.

Foreword

Climate change will not only determine the future of Europe; it is already part of our present. With devastating floods, droughts, wildfires and deadly heat waves, our continent is quickly learning about the gravity of this crisis. Moreover, science is telling us that this is merely the new normal.

If we take science seriously and want to minimise the disruption and destruction from the climate crisis, we need to act, and we need to do it now. That means keeping the planet's temperature rise to 1.5 degrees, as we committed to at COP26 in Glasgow. If we go beyond that, we will reach a number of tipping points. At that stage, the climate crisis would no longer be under our control and we then cease to be masters of our own destinies.

The European Union has set itself the goal of becoming the first climate-neutral continent on the globe by 2050. With the European Green Deal as our strategy towards climate neutrality, with the European Climate Law setting binding EU climate targets for 2030 and 2050 and with further legislation to implement it, we are equipped with the necessary framework to face this crisis in Europe.

Delivering on our targets and ambitions is not only a moral obligation to future generations and thus to the future of Europe, it is also a prerequisite for peace and prosperity in the world as a whole. Russia's brutal and unjustified aggression against Ukraine and the COVID-19 pandemic reminded us how fragile these are. It is our responsibility to build back better, to become more resilient and more sovereign in our energy choices—and avoid fossil fuels from being used as a political weapon against us.

For more than two centuries, carbon has been the basis of our wealth. It has brought great economic advances as well as an overreliance on fossil fuels, for which we are now paying the price. Turning this model around in just two decades is very difficult, but not impossible. We are making headway on all areas of the Green Deal—reducing greenhouse gas emissions, reversing biodiversity loss and advancing the circular economy.

When it comes to the success or failure of our transition, there are two elements to highlight. The first one is that it can only work if it happens in a fair and socially viable manner. This is going to be the decisive factor of our failure or success. For a peaceful and prosperous, climate-neutral Europe of tomorrow, we need to deliver on today's promise of 'leaving no one behind'.

After all, if you worry about making it to the end of the month, the end of the world is not your biggest concern. So, we need to address such insecurities and put people at the heart of this transition. This is why we set up a Social Climate Fund along with our climate policies and why we support regions in transition with the Just Transition Fund. Our transition towards climate neutrality by 2050 has to involve everyone and it has to be just. Otherwise, it just won't be.

Another risk we face is the dramatic loss of biodiversity. It is a direct threat to our survival and needs to be addressed with the same conviction as the climate crisis. Because both are two sides of the same coin: biodiversity loss is accelerated by climate change and at the same time exacerbates it. Fortunately, governments, businesses and the wider public are increasingly aware of the dangers posed by a looming ecocide.

Nature is our biggest ally in the fight against the climate crisis. We have to tackle both crises at the same time and ensure that our efforts to fight climate change help to reverse biodiversity loss—and vice versa. It is possible: when we protect and restore wetlands, peatlands, coastal and marine ecosystems, when we develop urban green spaces and install green roofs, when we manage forests and farmland in a sustainable way, we mitigate and adapt to climate change, but we also ensure clean water, healthy soils and space for nature to flourish. Instead of making nature pay the price for our pollution, we need to restore and protect nature so it can protect us.

As Europe, we need to show leadership and intensify our efforts with partners around the world to ensure a global leap in ambition—and deliver on that ambition. Furthermore, we need prove to the world that despite the pandemic and war, Europe stays the course. We remain on track, which is a socially just path towards climate neutrality.

In this decade, our overarching task is also to learn how to live within the boundaries of our planet. If we don't, Mother Earth will shed humanity as an old skin and simply get rid of all of us. This is not about saving the planet—the planet has done fine without us for billions of years. This is about saving humanity.

Time is running out, but with enough political will, we can still do it.

European Commission   Frans Timmermans
Brussels, Belgium

Keywords

Climate change, Future of Europe, Energy dependencies, EU climate policies, Public perception of the climate crisis

Why This Book?

A heat wave of over 40 °C hit Spain in May, the highest temperatures ever recorded on the Iberian Peninsula so early in the year. Around the same time, Great Britain also experienced record-breaking temperatures. Southern Europe was once again ravaged by forest fires which destroyed thousands of hectares of forest. Meanwhile, Germany's rivers have been drying up and Italy is enduring a state of emergency along the Po with a third of its harvest being threatened. According to the European Drought Observatory, large parts of southern Europe are significantly drier than normal with a consequent greater risk of fire and hence many areas are on high alert. The European Environment Agency is quoted as saying that 100 million people, almost one in seven inhabitants, are affected by water scarcity, which even includes the south of Sweden. These are not future scenarios for Europe, but current news.

This news also takes into account Russia's military invasion of Ukraine on 24 February 2022. Its implications touch literally all policy areas, including the fight against global warming. There are clear interdependences between climate change, the reduction of energy dependencies and food security, signalling that transformation of our economies has never been more visible, an issue which is rapidly gaining in importance and urgency. We are once more at a major turning point.

The climate crisis undoubtedly remains the greatest challenge of our time, despite Russia's war in Ukraine and the massive risks this poses regarding energy and food security. Widespread forest fires, landslides, floods, coastal degradations, increases in temperature and changes in the rainfall regime know no borders, no cultural differences, yet deepen societal cleavages within countries, not just across Europe, but world-wide. It determines our present and future, thus impacting each and every one of today's decisions and non-decisions.

United Nations Secretary-General António Guterres could not have been clearer when he told delegates at COP26, the November 2021 UN Climate Conference in Glasgow, that an addiction to fossil fuels is pushing humanity to the brink of extinction: 'We face a stark choice: Either we stop it—or it stops us. It's time to say: enough. Enough of brutalizing biodiversity. Enough of killing ourselves with carbon. Enough of treating nature like a toilet'. Nevertheless, the results of COP26 remained embarrassingly modest compared with the challenge. Gleams of hope, especially amongst young people, have been dimmed by a rather reserved level of political commitment from world leaders.

The European Union's (EU) answer to climate change is its Green Deal, setting out international environmental standards which will guide the Union on an ambitious journey to make Europe the first climate-neutral continent by 2050. Its climate goals and proposed package of binding legislation to transform our economies will pave the way towards climate neutrality and fast-track strategic autonomy. It is a complex and highly ambitious objective, but also extremely necessary as man-made disasters are occurring ever more frequently. Geopolitical tensions are impacting energy and food prices, supply chains are under pressure and migration is moving back into the headlines. If we are losing the fight for our climate, all those tensions will rise even further and the consequences will be catastrophic.

With the Russian war against Ukraine waging, energy security is now paramount. Short term, it may hinder green transition because of the necessity to: reopen coal power stations or extend their lifespan; invest in liquid natural gas; or recommit to nuclear energy and fossil fuels from other parts of the world. Yet, the costs of permanent wrongdoing are high. Whilst recent developments might block current efforts, nevertheless they will ultimately drive Europe towards more energy auton-omy based on renewables. War in Europe and the COVID-19 pandemic have not only brought cross-country dependencies to the surface, but they are also changing our mind-set and raising public awareness regarding the detrimental effects of fossil energy dependency, geopolitical power politics and the environmental crisis.

Climate change also provides opportunities to build new more resilient economic and social models. With European climate goals set, the initial starting positions and approaches to their achievement differ from country to country. Factors such as geography, history, science, politics and civil activism are likely to be some of the driving forces in this venture and the consequences of war in Ukraine will also be felt differently. Accordingly, authors from each of the respective 27 EU member states and 12 selected countries in the EU's neighbourhood analyse in short and concise contributions a whole variety of aspects regarding their respective countries' fight against climate change, including: its level of importance; factors determining a country's path towards climate neutrality; champions and opponents of the fight against climate change; possible solutions; and public perception of the climate crisis. Hence, there are 39 contributions expressing opinions, which provide a better under-standing of national differences and initial positions regarding the environmental challenge. Presented in this way, we reveal yet another kaleidoscope of diverse approaches regarding climate policies.

Despite the efforts of many, there is as yet no truly unified European approach, but rather a varying country-by-country effort, in the shape of different ambitions which give rise to unequal national climate and energy plans. Given the diverse national energy mixes, this is hardly surprising when one considers: Cyprus, for example, having a 91% dependency on fossil fuels; the important role of coal in Bulgaria, Poland, Slovenia and Slovakia; as well as many alternating geological possibilities and energy relations. For some this has led to a rethink in regard to phasing out coal, such as: North Macedonia, which was the first Western Balkan country to reach the decision to abandon coal, but then announced in spring 2022 the opening of two lignite mines; as well as Belgium, Finland, Romania and Slovenia, for whom the war in Ukraine has led to immediate extensions of nuclear power. The need for secure energy sources especially during cold winter months has emphasised the importance of nuclear power. This can be seen in Finland, for example, with the long overdue fifth reactor build by French Areva and German Siemens becoming fully operational by the end of 2022.

Most Europeans feel anxious about climate change. However, it frequently appears that the general public considers it to be a more serious problem than the political elites, as is the case in Czechia. Sometimes even national courts are pushing the green agenda. In spring 2021, for example, the German Federal Constitutional Court ruled that the Federal Climate Change Act of 2019 was insufficient to protect the freedoms of young people and future generations. Consequently, a revised Act now requires Germany by 2030 to have reduced its greenhouse gas emissions by 65% compared with 1990 and to have achieved greenhouse gas neutrality by 2045.

In many European countries, this challenge has led to a more active civil society, often in coalition with the scientific community, pushing political and economic actors into greening their political agenda. As environmental degradation tends to affect already vulnerable strata of society disproportionately, in demographic terms the climate movement's supporters tend to be younger, rather urban, mainly women and with a high level of qualifications. In Bosnia-Herzegovina, for example, one of the most prominent protests in recent years was named 'The Brave Women of Kruščica'. During 2017, many women from the small Bosnian village of Kruščica organised themselves and went on duty for more than 500 days, blocking the local bridge, preventing the passage of excavators and thus the construction of mini-hydro power plants on the river. Despite the authorities deciding to send in police who used physical force against activists, the women of Kruščica succeeded, with the Cantonal Court in the city of Novi Travnik annulling the relevant permits, thereby halting construction of the two planned mini hydropower plants. In other countries, such as Belgium, Austria, France, Germany, Greece, the Netherlands, Slovakia, Sweden and Switzerland, there have been predominantly 'youth for climate initiatives' mobilising public and political support for action.

The deployment of renewable energy projects (particularly large-scale wind and hydropower), though, has also been generating increasing opposition from some local communities and environmental non-governmental organisations. In Slovenia, for example, the so-called BANANAs (Build Absolutely Nothing Anywhere Near Anything) and NIMBYs (Not in My Backyard) groups attract some public support. In Czechia, the ideological narrative of both right-wing and left-wing governments has long emphasised the economy's primacy over the environment. Correspondingly, environmental activism has often been denigrated and the green movement has been pejoratively referred to as 'eco-terrorists'.

At the same time, a growing number of subnational authorities and cities have emerged as champions in tackling climate change. Through their climate agendas, many French cities, such as Grenoble, Lyon, Nantes, Poitiers, or regions such as Grand Est, Nouvelle Aquitaine and Val de Loire, are increasingly challenging the

State in its ability to develop a holistic and country-wide transition strategy. The same seems to hold for other cities, such as Helsinki and regions such as North-Rhine Westphalia.

Moreover, in other countries, climate change became a key topic in recent parliamentary and/or presidential elections, as was the case in Finland, France, Germany, Lithuania, Slovenia and the Netherlands. In Latvia and Lithuania, strong public support for environment movements dates back to the countries' historical heritage of resisting Soviet policies. The environmentalist movement in Lithuania criticised Soviet industrial practices as unsustainable, campaigning against expansion of the Ignalina nuclear power plant, which in the late 1980s operated two reactors of the same type as that at Chernobyl. This constituted a visible part of its independence movement. Today, the annual 'big clean up' in Latvia is a national event deeply rooted in society and also linked to the country's past Soviet oppression.

Some European countries, such as the Netherlands, Austria, Norway, Denmark and Iceland actively define themselves as frontrunners in the environmental agenda. An important part of Iceland's road towards climate neutrality, for example, is the sales ban of new fossil-fuel cars from 2030 and tax reductions for electric cars. This has already placed Iceland second behind Norway in terms of electric cars purchased per capita. For its part, Austria has successfully introduced a simple single climate ticket for all public transport in 2021. The 2020 Danish Climate Act sets a 70% greenhouse gas emissions reduction target by 2030 (compared with 1990 levels) and guarantees climate neutrality by 2050. Some countries, such as the United Kingdom (UK) and Ireland, had already declared a climate and biodiversity emergency in 2019. Others, such as Hungary, Poland and Czechia, are more like passengers on this journey and still need to be convinced by external factors before they will change course. According to the climate change performance index, Czechia, Hungary and Poland have even gone into reverse of late. In North Macedonia, the air pollution in 2018 and 2019 surpassed the ceiling for air pollutants determined by national emission reductions plans (sulphur dioxide and dust, in particular) and resulted in the opening of a case against North Macedonia (and Kosovo) by the Vienna-based Energy Community Secretariat. It seems that where short-term political considerations and cost–benefit analysis prevail, sustainable medium- to long-term investments suffer. Finding the right balance between the two is key, not only through environmental and technological know-how being shared, but also by political awareness about the cost of non-action being raised.

However, there is also a crucial social component to this green transition. While an active climate policy might enjoy substantial abstract majoritarian political support, the public mood may still change once individual social costs come to the surface and concrete restrictions to levels of consumption are directly experienced, with topical referenda in Switzerland being a case in point. By contrast, half of the households in Romania still use wood for heating, the majority being in rural areas where alternative heating sources are lacking and therefore energy poverty is widespread. A particular social, rural and urban divide is also being witnessed. In Bulgaria, for instance, the 'Marica Iztok' coal mining and energy complex employs around 10,000 people directly and 20,000 in related industries. In the Polish coal region of Śląsk, 100,000 people are still employed in the coal industry (mainly power plants and mines).

As poorer families struggle to meet their energy and food bills, pundits will be eager to argue that city-based, middle-class obsessions with climate change should not be given priority over pressing present economic needs fuelling Eurosceptic sentiments, in other words the green energy revolution is being 'forced by Brussels'. Hence, social dimensions of the green agenda and financial assistance for the more vulnerable parts of society are essential for sustainable and broad public support as well as a stable political environment. It comes as no surprise, therefore, to find that many countries would rather argue for raising investment into research and innovation than discipline behaviour.

For those states which lack economic resources, international and European financial support to achieve the Paris climate goals acts as important leverage. Direct allocation of funds to recognised active civil society organisations or local governance levels as well as a stronger green conditionality would also be of tremendous help. In Cyprus, for example, it is the Cyprus-EU Partnership Agreement and the Just Transition Fund, which are part of the Union's Cohesion Policy that include the European Regional Development Fund, the European Social Fund Plus and the Cohesion Fund providing for investment of over EUR 387 million in energy efficiency, renewables and reduction of greenhouse gas emissions. Furthermore, the European Bank for Reconstruction and Development has been active in Turkey since 2010 and has created financing facilities for various sectors, such as the Turkey Sustainable Energy Financing Facility, the Mid-size Sustainable Energy Financing Facility and the Turkish Residential Energy Efficiency Financing Facility. Ultimately, EU funds, including those related to the national Recovery and Resilience plans, will continue to help in the prioritising of green spending.

When knowledge about climate change is vague and the EU's goal of becoming a global leader in the fight against climate change appears abstract, people can quickly conclude that the problem is very distant. Societies, such as those in Bulgaria, Czechia, Poland or Slovenia, could therefore derive great benefit from information and educational campaigns which raise environmental awareness and directly impact the political agenda. In countries such as Portugal, where public interest is mostly rooted in a persisting ruralist sensibility, the deepening role of environmental education, notably in schools, and rapidly increasing media coverage could be of additional value. The current inflationary environment, especially in regard to high fossil fuel prices resulting from global mismatch between supply and demand as well as the war against Ukraine, creates strong incentives for the population and businesses to change consumption patterns and search for ways to save energy. State institutions should facilitate this process by providing information and shifting towards more sustainable practices. Other countries, such as Sweden, have set up administrative bodies to supervise effective political implementation of national plans.

As pollution does not care about national borders, more European and international cooperation is needed to counter the green governance problem. The fragmentation of climate competences is not only a test at the level of Member States, say in consociationalist Belgium due to its three regions, but also very much a European multi-level challenge. Yet, where there is a single market, better energy interconnectedness and one single climate policy is a *sine qua non* in realising Europe's full potential and pushing for European energy solidarity. Joint energy procurement is but one step in this direction; sooner rather than later we will need a coordinated, long-term approach to a European super-grid integrating the level of renewables.

All countries face unique and exceptional circumstances to which policy-making has to adapt. Thinking 'outside of the box' and helping to improve people's understanding of these issues might be a cumbersome exercise, yet it is important not to focus on limitations, but on new possibilities, best practices and joint solutions. This holds true not only for European islands such as Cyprus, Malta, Ireland or Iceland, but it is also a matter of survival, for example in the Dutch context. Financial frugality in many EU capitals, say Copenhagen and Helsinki, poses risks to an effective European economic energy transition and may even raise costs in the medium to long term. Strategic sustainable long-term planning would be an asset. However, a reluctance to accept that green priorities must at times off-set frugal issues could inhibit the chances of advancing climate polices at European level. Starting points differ, but to fulfil our joint cross-border objectives we have to enable a just transition for everyone.

If President Macron's idea of a European Political Community is to prevail, it needs to embrace climate change as one of its key components. Our volume helps by covering European countries with diverse levels of integration and cooperation. It analyses the different EU and European experiences and approaches to climate change, the way it is perceived by its people and its overall importance for the future of Europe. This book also addresses an audience far beyond the typical academic niche interested in European politics. It is rather a guidebook in our 'Views from the Capitals' series taking us through a tremendously varying and exciting political landscape of Europe that is constantly changing. Its countries constitute the individual and unique pieces of a puzzle, which together reveal a bigger European picture. As a guidebook, it favours lexical purpose as much as comprehensive comparative reading. Students and teachers may find a score of questions, differences and common ground to explore more deeply in seminar papers and theses. Practitioners will benefit from the short overviews being presented on possibly the biggest topical issue of our times, and, for all of us who are interested readers, it demonstrates the breathtaking diversity that sometimes divides, but ultimately unites and defines this continent.

We are particularly grateful to Eva Ribera, Project Officer at the Trans European Policy Studies Association, for her editorial processing and tireless efforts in making this project become a reality.

July 2022   Michael Kaeding
                                                             Johannes Pollak
                                                               Paul Schmidt

Contents

Contributors

Tinatin Akhvlediani PMC Research Center, Tbilisi, Georgia

Katrin Auel Institute for Advanced Studies and Austrian Society for European Politics, Vienna, Austria

Margherita Bianchi Istituto Affari Internazionale (IAI), Rome, Italy

Ana-Maria Boromisa The Institute for Development and International Relations (IRMO), Zagreb, Croatia

Karlis Bukovskis Latvian Institute of International Affairs (LIIA), Riga, Latvia

Luke O'Callaghan-White Institute of International and European Affairs (IIEA), Dublin, Ireland

Svitlana Chekunova Razumkov Centre, Kyiv, Ukraine

Jens Mattias Clausen Think Tank Europa & Concito, Copenhagen, Denmark

Danijel Crnčec Centre for International Relations - University of Ljubljana, Ljubljana, Slovenia

Brendan Donnelly The Federal Trust, London, UK

Joanna Dyduch Jagellonian University, Kraków, Poland

Vedran Dzihic Austrian Institute for International Affairs, Wien, Austria

Christian Frommelt Liechtenstein Institute, Gamprin, Liechtenstein

Magdalena Góra Jagellonian University, Kraków, Poland

Christina Goßner IEP Berlin, Berlin, Germany

Charlotte Halpern Science Po, Paris, France

Mark Harwood Institute of European Studies, University of Malta, Msida, Malta

Gunilla Herolf Swedish Institute of International Affairs, Stockholm, Sweden

Edmond Hoxha Polytechnic University of Tirana, Tirana, Albania

Danijela Jacimovic Faculty of Economics of University of Montenegro, Podgorica, Montenegro

Juha Jokela Finnish Institute of International Affairs, Helsinki, Finland

Giorgos Kentas University of Nicosia - Cyprus Centre for International Affairs, Nicosia, Cyprus

Olena Korohodova Faculty of Economics of University of Montenegro, Podgorica, Montenegro

Petr Kratochvíl Institute of International Relations Prague (IIR), Prague, Czechia

Guido Lessing Luxembourg Centre for Contemporary and Digital History, Esch-sur-Alzette, Luxembourg

Gunnhildur Lily Magnusdottir Institute of International Affairs, Reykjavík, Iceland

Niklas Mayer Maastricht University, Maastricht, Netherlands

Dániel Muth Institute of World Economics (IWE), Centre for Economics and Regional Studies, Budapest, Hungary

Ivan Nachev New Bulgarian University, Sofia, Bulgaria

Aleksandra Palkova Latvian Institute of International Affairs (LIIA), Riga, Latvia

Hristo Panchugov New Bulgarian University, Sofia, Bulgaria

Irena Rajchinovska Pandeva Facutly of Law Iustinianus Primus, Skopje, North Macedonia

Angeliki Papantoniou Queen Mary University of London, London, Greece

François Roux Egmont, Brussels, Belgium

Ezgi Ediboğlu Sakowsky Middle East Technical University (METU), Ankara, Turkey
Sabanci University, Istanbul, Turkey

Frank Schimmelfennig ETH Zurich, Zurich, Switzerland

Luisa Schmidt Institute of Social Sciences - University of Lisbon, Lisboa, Portugal

Paul Schmidt Institute for Advanced Studies and Austrian Society for European Politics, Vienna, Austria

Mihai Sebe European Institute of Romania, Bucharest, Romania

Ditte Brasso Sørensen Think Tank Europa & Concito, Copenhagen, Denmark

Natasza Styczyńska Jagellonian University, Kraków, Poland

Zdeněk Sychra Institute of International Relations Prague (IIR), Prague, Czechia

John Szabo Institute of World Economics (IWE), Centre for Economics and Regional Studies, Budapest, Hungary

Kacper Szulecki Norwegian Institute of International Affairs (NUPI), Oslo, Norway

Funda Tekin IEP Berlin, Berlin, Germany

Lara Lázaro-Touza Energy and Climate Programme, Elcano Royal Institute, Madrid, Spain
Business Administration, Centro de Enseñanza Superior Cardenal Cisneros (attached to Universidad Complutense de Madrid), Madrid, Spain

Eliza Vaş European Institute of Romania, Bucharest, Romania

Viljar Veebel Estonian Foreign Policy Institute, Tallinn, Estonia

Ramūnas Vilpišauskas Vilnius University, Vilnius, Lithuania

Donald Wertlen Comenius University Bratislava, Bratislava, Slovakia

John Szabo is none of World Economics (WE) Centre for Economics and Regional Studies, Budapest, Hungary

Part I

Member States

Austria: Weathering the Storm and Greening the Economy?

Katrin Auel and Paul Schmidt

Austria still has a long way to go in its efforts to achieve both the European climate targets and the even more ambitious self-set objective of climate neutrality by 2040. According to the Austrian Court of Audit, greenhouse gas emissions decreased by 24% on average in the EU over the last 30 years (from 1990 to 2017), while they increased by 5% in Austria. Austria was one of six EU Member States that had failed to reduce carbon emissions. This situation has not improved much since then, although political ambition to green the economy is growing. Yet, in 2021 CO_2-emissions had already reached pre-pandemic levels, making the 2020 decrease a possible 'one-time outlier' resulting from pandemic-related issues. Outside of the Emission Trading System, the main culprit is transport, alone responsible for around 47% of greenhouse gas emissions in 2018. Key industries with the largest carbon footprint are found especially in the steel sector, but also other sectors such as oil, paper, chemistry, energy and construction materials. As a medium-sized, export-oriented country with a strong industrial base together with a large share of small and medium-sized enterprises, as well as a high level of natural gas dependency on Russia, for Austria climate change is thus both a highly salient and challenging topic.

Within the EU Recovery and Resilience Facility, which aims at making European economies and societies more sustainable, resilient as well as better prepared for green and digital transitions, Austria has committed to 32 investments and 27 reforms. These are supported by EUR 3.46 billion in EU grants of which 59% are flagged for use on climate objectives, investing, for example in sustainable mobility with zero-emission transport and the expansion of electrified trans-European rail networks, including links to regional lines. Companies' investment in low-emission buildings and vehicles will be supported, as will the phase-out

K. Auel (✉) · P. Schmidt
Institute for Advanced Studies and Austrian Society for European Politics, Vienna, Austria
e-mail: auel@ihs.ac.at; paul.schmidt@oegfe.at

3

of oil and gas heating in private homes. Further investments include a biodiversity fund and the recycling of beverage containers.

The government has recently implemented further reforms complementing these measures, such as the Renewable Energy Expansion Act (2021) which provides EUR 1 billion annually in subsidies for investments with the aim of: reaching full electricity generation from renewables by 2030; introducing the so-called 'climate ticket', a flat rate season ticket for public transport that can be used throughout Austria; as well as establishing a soil protection strategy. In addition, a carbon levy was introduced in October 2021. Originally planned for July 2022, then postponed to at least October 2022 due to high inflationary pressure, this levy will increase from an initial EUR 30 per ton of CO_2—or equivalents—to EUR 55 per ton in 2025.

Given that costs are to be passed directly on to consumers to achieve behavioural change, the government has put in place corrective social measures, for example in the form of a so-called 'climate bonus' and tax reliefs for those on lower incomes. This strategy also includes an anti-carbon leakage regulation aimed at preventing Austrian production from relocating to less regulated countries. This includes a refund of 65–95% of the levy to emission intense companies competing internationally, with the stipulation that these funds must be used predominantly for climate protection. Regrettably, at least for now other important legislative initiatives and processes seem to have ground to a halt. This is true for a list of climate-harming subsidies, the abolition of which was already decided under the Austrian People's Party (ÖVP)—Freedom Party of Austria government in 2018, but still outstanding by Spring 2022. This has also affected renewal of the national Climate Protection Act, the previous 2011 Act having expired in 2020.

Climate experts have repeatedly warned that policies implemented so far are barely enough and whilst the goal of carbon neutrality by 2040 is laudable, this target is in essence no longer achievable. Regrettably, the government's powerful finance ministry and chancellery, both led by the conservative ÖVP, are considered by some to be the strongest procrastinators. Clearly, it must be borne in mind that the current ÖVP-Greens government suffered the disadvantage of being confronted with the COVID-19 pandemic almost immediately after taking office in January 2020. Yet the Greens, who are in charge of the climate ministry, are criticised as being 'caught in the ÖVP's symbolic climate politics', which argues that steps to fight climate change needs to be made 'with common sense and reason', relying mainly on innovation and new technologies rather than abstinence. Faced with additional worries due to the Russian war against Ukraine, which has produced massive inflationary pressure and challenges to energy security given Austria's high level of gas dependency on Russia, the current chancellor, Karl Nehammer (ÖVP), has not been able to make climate policy a top priority since coming into office early in December 2021. To reduce energy dependence and prepare for a complete cut-off from Russian energy imports, Austria's emergency plans now also include reactivation of its highly polluting coal-fired power station in Styria.

At the same time, surveys clearly indicate that most Austrians consider climate change to be a serious problem, with only one-third of respondents believing that the government is already doing enough to combat the crisis. While a majority see responsibility for climate action resting with the government, business or the EU, close to half Austrian residents also accept that they are personally responsible for tackling climate change and indeed this commitment to individual behavioural changes would appear to be above the EU average. Unsurprisingly, though, private action focuses mainly on small steps such as avoiding excess packaging or recycling waste, while larger changes are met with more reluctance. The climate ticket is a popular initiative, yet less than half believe that public transport must become climate friendly. Only around a third are prepared to: support the proposed internal combustion engine ban from 2030, minimise the use of private cars, and avoid air travel for holidays. Although a majority think that the government is not doing enough, it seems that most people support a careful, common-sense approach to climate policy that would not hurt too much. Austrians also seem largely confident that the government will undertake the necessary political measures to combat climate change, at least within the next 5–15 years.

There is, though, also a sizeable minority who doubt that the government will ever be able to implement enough necessary measures. It is against this background that a national citizens' initiative organised in June 2020 and—despite the COVID-19 pandemic—backed by 380,000 signatures prompted parliament to request the government to set up a citizens' assembly. This assembly was initiated in January 2022, tasked with the deliberation and elaboration of policy measures aimed at making Austria carbon neutral by 2040.

Recommendations

At this point, with the Russian invasion of Ukraine wreaking havoc, it seems unlikely that the joint fight against climate change will form the basis of a new narrative for European integration and cooperation in Austria. Austrians are at best pragmatic Europeans and politicians have regularly fuelled public anti-EU sentiment by blaming 'Brussels' for unpopular policies. However, the need for rapid reduction in energy dependency on Russian fossil fuels could also speed up a greening of the economy. Nevertheless, whilst the economic and social costs of this transformation will be substantial and need to be well balanced within society, taking no action at all would ultimately be even more expensive. It seems likely that populist politics will seize this opportunity to continue the well-known blame game in which more drastic measures adopted for the fight against climate change are blamed for price hikes and related to aloof EU legal obligations. Yet, Austria is no 'island of the blessed' that can simply close its eyes and decouple from a global climate crisis, not to mention the economic war against Russia. The majority of its citizens are well aware that contamination knows no borders and tackling climate change needs to be a priority. To weather the storm, generationally sensitive and responsible politicians need to build on this and make the case for cross-border, joint European action.

Katrin Auel is Head of the Research Group European Governance, Public Finance and Labour Markets at the Institute for Advanced Studies (IHS) Vienna, which she joined in 2012. After completing her studies at the University of Konstanz, she held positions at the University of Halle-Wittenberg, the University of Hagen, the University of Oxford and the European University Viadrina. Her research focuses on Europeanisation and legislative studies.

IHS is an independent research institute covering the areas of Economics, Political Science and Sociology and a member of TEPSA.

Paul Schmidt is Secretary General of the Austrian Society for European Politics, which promotes analysis and communication on European affairs. Prior to that he worked at the Oesterreichische Nationalbank, both in Vienna and at their Representative Office in Brussels at the Permanent Representation of Austria to the European Union. Paul currently dedicates his time to discussing and advancing European integration, with a special focus on the future of the EU and public opinion. His analyses, comments and op-eds are regularly published in Austrian and international media.

The Austrian Society for European Politics (Österreichische Gesellschaft für Europapolitik) was founded in 1991 and aims to promote and support communication and analysis of European affairs in Austria. With its headquarters in Vienna, the Society is a non-governmental and non-partisan platform mainly constituted by the Austrian Social Partners and the Oesterreichische Nationalbank. The Society is also a member of TEPSA.

Belgium's Climate Policy: High Expectations, Low Performance

François Roux

Heightening Public Awareness

Belgian awareness of climate change has been the subject of regular public surveys since 2005. The fifth edition of this survey, published in March 2022, reveals that climate change is now the main concern for eight out of ten Belgians, compared to seven out of ten Belgians in 2009. Wide dissemination of information on climate change, particularly the Intergovernmental Panel on Climate Change reports, through both traditional media and the Internet, explains why only 8% of Belgian citizens say they are not concerned. The occurrence of spectacular disasters, such as the floods in July 2021, when record rainfall in Belgium, Netherlands and Germany caused many rivers to burst their banks provoking many casualties and damages, has also raised awareness and rekindled activism amongst the younger generation.

Low Performance

According to the 'Climate Change Performance Index' published annually at the UN Climate Change Conference, Belgium's efforts have fallen from 16th place in 2015 to 49th place by the end of 2021. Belgium has not seen any reduction in greenhouse gas emissions as a result of the 2014–2019 legislature which was led by a coalition government comprising 'Mouvement Réformateur' (Francophone liberals) and the 'Nieuwe Vlaamse Alliantie' (Flemish nationalists). Reductions observed in 2020 were ascribed to the COVID-19 crisis and not to any structural action by the federal and regional governments. Yet a recent study commissioned by the Federal

F. Roux (✉)
Egmont, Brussels, Belgium
e-mail: f.roux@egmontinstitute.be

M. Kaeding et al. (eds.), *Climate Change and the Future of Europe*, The Future of Europe, https://doi.org/10.1007/978-3-031-23328-9_2

government showed that a climate neutral Belgium by 2050 is achievable and the government has therefore subscribed to the 55% reduction in greenhouse gas by 2030.

Energy Vulnerability

Some 80% of the energy consumed in Belgium comes from fossil fuel resources. More than 90% of our energy comes from abroad. Before the war in Ukraine, Russia provided 30% of the oil, 25% of the uranium and 6% of the natural gas consumed in Belgium. Within Belgium's electricity market, the energy mix included 20% renewables, 30% fossil fuels and close to 50% nuclear power in 2021. Despite the 2025 nuclear phase-out agreed in October 2020, the current government (a coalition of seven different political parties) decided on 18 March 2022, to extend the useful life of two relatively new plants in Belgium (Doel four and Tihange three) until 2035.

This decision, together with the construction of two gas-fired power plants, should ensure the country's electricity supply. The European Commission will need to approve Belgian state financing and market terms for these new electricity providers before they enter into force in 2025. Given the geographical and political constraints renewables are facing on land, Belgium is turning to wind energy projects in the North Sea, expected to be operational by 2030, and is also engaged in long-term research into the use of hydrogen.

The Involvement of Companies in the Transition

Energy-intensive industries (the chemical industry in the Antwerp region, the special steel industry in the Ghent region and the glass and cement industries in Wallonia) have a key role in supporting the transition towards climate neutrality. According to a study conducted by PricewaterhouseCoopers, just under a third (28%) of energy-intensive industries have already implemented the necessary measures to reach the zero-emission objective by 2030. Over the last two decades, sectors such as waste management, electricity production and manufacturing have significantly reduced their greenhouse gas emissions. However, road transport (20% of emissions) and buildings (19% of emissions) remain problematic and are currently preventing further reductions of overall emissions in Belgium.

Pros and Cons in the Fight Against Climate Change

On the side of climate champions, first we find civil society, including the 'climate coalition' which comprises about 90 Belgian associations (bilingual, French and Dutch speaking) divided between environmental organisations and social/socio-cultural associations. Amongst political parties, the Greens (ECOLO/Groen) form part of the current federal government's coalition and are also featured in the regional governments. On the opposition side, there is no climate sceptic party in Belgium. However, the fragmentation of climate competences between three regions, three language communities and the federal government poses a fundamental problem for governance. The complexity of Belgian procedures and legislative tools makes it impossible to meet the European Commission's requirements, which has repeatedly pointed out the lack of consistency in Belgium's national plans. These climate efforts are merely a compilation of regional strategies and do not represent an integrated and systemic vision of a green transition at country level. Finally, most stakeholders perceive the current mechanisms of climate policy dialogue to be opaque and undemocratic.

Public Perception of Government Action

According to a European Investment Bank survey, 75% of Belgians believe that they are more concerned about the climate emergency than their own government. More than half of Belgian citizens in the March 2022 opinion poll on climate change indicated that they would certainly consider the different political parties' climate views before voting in the next elections. Citizens' critical views reflect the distance between an anxious population and a government struggling to meet the European climate objectives. This discontent led to success for the 'Youth for Climate' movement in Belgium, launched in 2019 by two young students Anna De Wever and Kyra Gantois, having been inspired by Greta Thunberg. Following the protests and climate marches, civil society joined forces with academia to promote a 'climate law' which was intended to fix the weaknesses in Belgian climate governance. Regrettably, this was ultimately not accepted by the Belgian parliament.

Belgium's Position in the European Context

Two out of three Belgians believe that the EU should play a key role in taking measures to combat climate change. According to the same survey, 44% even believe that Belgium should take the lead in this area. Belgian stakeholders are aware that EU action could be globally influential on climate change. They expect the Union to achieve fair transition to a sustainable economic model through: (1) a new industrial policy focused on innovation, which would preserve a level playing field for companies including small and medium-sized enterprises; (2) a European framework that would allow the preservation of fair and inclusive growth thanks to a carbon border adjustment mechanism to avoid carbon leakage; as well as (3) a fluid and efficient Emissions Trading System.

Recommendations

Unlike its possible negative effects on biodiversity and the European agricultural sector, the Ukrainian crisis could ironically help promote the Green Deal's success on a European scale by accelerating European integration in the energy sector.

To take an active part in shaping EU policy and implement the Green Deal objectives, Belgium must: improve its ability to govern directly on climate issues; dedicate significant financial support to the development as well as implementation of necessary technological breakthroughs; and impose real energy sobriety that is tempered by consideration for disadvantaged populations.

Despite its handicaps (flat, densely populated country with a 65 km coastline), Belgium should make best use of its funds—including the European Resilience and Recovery support—to finance huge programmes for improving the insulation of buildings and reduce road transport.

François Roux is Senior Advisor at Egmont. Former Chief of staff of the President of the European Council (2019–2020), Permanent Representative of Belgium to the European Union and Sherpa of Prime Minister Charles Michel (2016–2019). Ambassador Roux was Director General for European Affairs at the Belgian Ministry of Foreign Affairs (2012–2016). He holds a master's degree in economics from the University of Paris and a master's degree in public Affairs and International Relations from the University of Louvain.

Egmont—the Royal Institute for International Relations—is an independent think-tank based in Brussels. Its interdisciplinary research is conducted in a spirit of total academic freedom. Drawing on the expertise of its own research fellows, as well as that of external specialists, both Belgian and foreign, it provides analysis and policy options that are meant to be as operational as possible. It is also a member of TEPSA.

Between Fighting for Climate Change and Fighting for Coal: The Bulgarian Case

Ivan Nachev and Hristo Panchugov

The Bulgarian economy's structure is focused on agriculture and tourism, thus making it particularly vulnerable to the effects of climate change. Yet the European Green Deal and any related decisions have generated only a relatively low level of interest in national political discourse. Neither climate change nor the aim of achieving a carbon neutral economy command any significant attention in political and scientific fields. As with many other topics, the Green Deal's accompanying financial support is reviewed in purely instrumental terms, with a focus which gravitates more towards the value and distribution of public funds related to its implementation in Bulgaria. Thus, it has been the subject of many high- and low-profile scandals related to political corruption and conflicts of interests regarding the allocation of these funds to companies that are closely linked to the government.

The lack of relevant information and clear political positions on any issues related to European green policies, such as carbon neutrality, recycling and sustainable use of resources, has created a gap between EU policy and Bulgaria's domestic political agenda. The EU's goal of becoming a global leader in the fight against climate change has remained abstract, running parallel with and sometimes against mainstream political issues. Whilst recently the main political parties have recognised this Green Deal as part of the EU Recovery and Resilience Plan for Bulgaria, nevertheless the topic is reviewed in terms of financial benefits only, in light of the funds that will be available for Bulgaria. The one party which is likely to push the green agenda forward will be the 'Green Movement', a member of the coalition 'Democratic Bulgaria'. However, still their agenda is dominated by local environmental issues, rather than EU Green Deal goals.

As much as the topic has found its way into domestic politics at all, it has been focused on the expected effects that the Green Deal will have on the national energy

I. Nachev · H. Panchugov (✉)
New Bulgarian University, Sofia, Bulgaria

sector. In this regard discourse has clearly been dominated by populist concerns over closing of the Bulgarian coal-fuelled power plants, clustered in the region of Stara Zagora, where they have been the biggest source of employment in recent decades. In the past 15 years over EUR 1 billion have been invested in modernisation of the Marica Iztok coal mining and energy complex. The complex has around 2 billion tons of coal deposits and employs around 10,000 people directly and 20,000 in related industries. The narrative of unemployment and economic stagnation along with the unavoidable rise in the price of electricity, has established itself firmly within regional and national policies. As a result, instead of active communication and preparation, there has been a general reluctance in political circles to discuss those issues openly, which has left room for bottom-up pressure to keep the power plans running to increase over the last couple of months, following approval of the Green Deal. As an illustration, in 2020 coal-produced energy formed 34.5% of Bulgaria's total energy mix, with renewables producing 23.8% and biomass 40%.

Not unsurprisingly, popular demand, fostered by economic and private interests, has led to a moratorium being adopted on the achievement of greenhouse gas emissions targets, thereby extending the life of coal-fuelled power plants until 2040. This will guarantee national energy independence and provide enough time for the negotiation of sufficient EU funds to allow for implementation of innovative technologies in low carbon energy production.

Following pressure from interested parties, recent Bulgarian governments have failed to establish a clear direction and timeframe for implementation of the country's Green Deal goals. Thus, it seems unfortunate that one of the major EU policy areas cannot find its way into the policy programmes and agenda of the Bulgarian political parties and hence government priorities, especially given the public opinion polls concerning climate change. A survey conducted in 2021 found that 85% of Bulgarians believe that climate change will have a significant impact over their lives and 55% approve the EU goals regarding reduction of greenhouse gas emissions. Furthermore, 70% believe that the fight against global warming should be the key priority for all Bulgarian political parties.

To a certain extent, these issues have found their way into the Bulgarian Recovery and Resilience Plan with EUR 1.7 billion having been dedicated to: the gradual phasing-out of coal; achieving a 40% reduction in power sector greenhouse gas emissions by 2025; and tripling power generation from renewables by 2026. However, as yet this has been determined on the basis that no coal-power plants will be closed. A substantial portion of the funds (EUR 924 million) will be dedicated to achieving energy efficiency in private and public buildings and EUR 666 million will be spent on sustainable transport. An official estimation claims that 59% of the plan's total allocation for reforms and investments supports climate objectives; however, a number of questions related to the specific projects and their accessibility to small and medium businesses remain unanswered. Given the track record of such measures in previous programme periods, the overall effectiveness of the projects targeted at achieving energy efficiency in private and public buildings is under serious doubt.

This data points to a severe clash between public and private interests in the Bulgarian energy sector, leaving the country side-lined in terms of the main EU political debate. This is yet another example that shows Bulgarian governments' lack of interest, expertise and negotiation skills within the complex EU negotiation process. This approach not only diminishes Bulgaria's bargaining potential, but also creates space for spreading EU scepticism.

Recommendations

Given the existing gap between public perceptions and political discourse, the EU should focus on separating the Bulgarian government from links with the coal-related interested parties. Targeted and well-designed incentives as well as pressure to develop new green energy capabilities in general and a clear strategy for the Stara Zagora region's economic development should be applied. Currently, there are three new large energy projects that meet the EU's recommendations on greener energy power, but they are still held back by concerns related to that region.

Mindful of public attitudes on this topic, Bulgaria's citizens can be serious and powerful allies in the quest to achieve the Green Deal's short- and mid-term goals. A deliberate communication strategy targeted at motivating and activating Bulgarians to pressurise the government and support initiatives in that direction may go a long way in severing ties with the coal industry.

Trading with greenhouse gas emissions remains an incomprehensible instrument, largely used as a populist means of explaining the rise of energy prices and the bad outcomes of EU green policy. Further focus on explaining consumer benefits should be developed.

The EU Commission should focus on evaluating the effectiveness and efficiency of any proposed projects, especially with regard to the energy efficiency of buildings. The design of these measures and their end results not only need to be transparent, but a clear link to EU climate goals must also be kept throughout this period. The current approach based on post-implementation evaluation is ineffective and focuses on controlling the spending, rather than the projects' effectiveness, thereby widening the gap between agenda goals and project implementation.

Ivan Nachev is a Bulgarian political scientist and an expert on political integration of the European Union. His interests are in the fields of political theory and practice, European values, European integration theories, strategies and political practices. He is a member of the Bulgarian Association for Political Sciences, the Institute for Public Policies and Partnership, the European Community Studies Association (ECSA) and Team Europe at the European Commission.

The New Bulgarian University was established in 1991 following a Bulgarian Parliament resolution. Its mission is to be an autonomous liberal education institution dedicated to the advancement of university education by offering accessible and affordable opportunities for interdisciplinary and specialised education as well as high-quality research. The University is also a member of TEPSA.

Hristo Panchugov is an Assistant Professor at the Department of Political Science at the New Bulgarian University. He is a graduate of the Central European University (Hungary).

The New Bulgarian University was established in 1991 following a Bulgarian Parliament resolution. Its mission is to be an autonomous liberal education institution dedicated to the advancement of university education by offering accessible and affordable opportunities for interdisciplinary and specialised education as well as high-quality research. The University is also a member of TEPSA.

Croatia: Needs Versus Capacity: Mind the Gap!

Ana-Maria Boromisa

Among the EU Member States, only Latvia and Romania are more vulnerable to climate risks than Croatia, which is itself being significantly impacted. The Czech Republic, Hungary and Croatia are experiencing the highest share of damage from extreme weather and climate events in relation to GDP. However, despite these worrying signs, implementing climate change mitigation and adaptation measures is not yet recognised as a priority in Croatia.

In the period 1980–2020, there were around 900 fatalities, with damage caused by weather and climate-related extreme events in Croatia estimated at EUR 2.860 million, with only 3% of losses having been insured. The danger from adverse impacts is continuing to increase as the frequency and severity of unusual events intensify. In 2021, the State Hydrometeorological Service reported to the World Meteorological Organisation 18 events which breached natural climate variability expectations, namely winds, heat waves, cold snaps, thunderstorms and heavy rains. For instance, on a single summer's day in June 2021, wind accompanied with hailstones in an area with about 20,000 inhabitants close to the city of Požega caused damage totalling many EUR millions. Unusual events, related losses as well as damage to people and nature show that climate change mitigation and adaptation is not only relevant but can also generate many benefits. However, the importance of climate action is not perceived as a priority either in government policies or public opinion. On the one hand, the share of citizens who consider that climate change is caused by human activities is relatively low and on the other hand, people consider that it is already too late to reverse the effects of climate change.

Only 4% of Croatian citizens consider environment and climate change to be amongst the most crucial issues being faced by the country, compared with 16% across the EU. Croatian citizens consider inflation (rising prices/cost of living),

A.-M. Boromisa (✉)
The Institute for Development and International Relations (IRMO), Zagreb, Croatia
e-mail: anamaria@irmo.hr

health, economic situation, unemployment, energy supply and immigration to be far more pressing. National strategies recognise the need to foster green and digital transition but have yet to implement any plan which sets Croatia on the pathway toward climate neutrality. The National Development Strategy has established the 2030 goal of 35% greenhouse gas emissions reduction compared to 1990. This is well below the 55% target required by European Climate Law. Furthermore, the Low-Carbon Development Strategy considers that the climate neutral scenario is for information only. It claims that climate neutrality is not feasible for Croatia by 2050, but that the country can achieve up to 44.8% emissions reduction by 2030 and up to 89.4% by 2050. The financing gap is identified as the main obstacle for achieving climate neutrality by 2050.

Croatia relies heavily on EU funding for investments in general and for climate action in particular. About 40% of grants worth EUR 6.3 billion available under the Croatian Recovery and Resilience Plan are being allocated to support climate objectives. This includes: (1) investments of EUR 789 million in energy efficiency and post-earthquake reconstruction of buildings; (2) EUR 728 million in sustainable mobility, notably in upgrading railway lines, autonomous electric taxis with supporting infrastructure adapted for people with disabilities, installing charging stations for electric vehicles as well as introducing zero-emission vehicles and vessels. In addition, the plan allocates EUR 658 million to low-carbon energy transition through modernising energy infrastructure, supporting investments for the production of advanced biofuels and renewable hydrogen as well as financing innovative carbon capture and storage projects. EUR 542 million will be invested in supporting businesses for green transition and energy efficiency, supporting projects aimed at boosting the green economy, sustainable tourism and investing in green technologies.

Tourism, with almost 20% contribution to national gross domestic product (by far the largest share in the EU), is a key sector within the Croatian economy. It is characterised with stays concentrated in coastal areas over the summer months. Increasing the resilience and sustainability of this sector requires infrastructure investments, inter alia for charging electric vehicles and waste management facilities. The new strategy for sustainable tourism is trying to address these issues. However, as the current 'sea and sun' tourism model suffers from a lack of skilled workers, providing more services might be challenging.

Lack of capacity that goes beyond financial constraints tends to be underestimated in policy documents. This includes a lack of suitably skilled human resources not only in the private sector, but also in ministries for policy making and regulatory expertise. Furthermore, there is a shortage of technical knowledge and managerial skills for project preparation along with the implementation of complex and lengthy administrative procedures. A combination of public administration fragmentation, an inefficient judiciary and corruption militate against facilitating necessary mitigation and adaptation investment.

EU policy guidance has enabled identification of the key sectors that are likely to be most affected with climate change and hence any necessary reforms and investments. However, the policy framework is very fragmented, with various aspects of risk reduction being dealt with in many different sectoral laws. In certain areas, such as forest fire prevention and forest fire fighting, Croatia's approach is widely recognised as the golden standard for the whole of Europe. However, this model is not even used in streamlining and integrating disaster risk reduction procedures as well as improving co-ordination among various actors. Whilst the improvement of co-ordination mechanisms among institutions has been included in country-specific recommendations, Croatia has made only limited progress in this area since 2019.

Recommendations

The Government should start systematically removing constraints for climate action. This includes improving awareness of climate risks and their consequences, such as costs emerging from non-action, integrating adaptation and mitigation measures in relevant policies as well as addressing capacity constraints. To address financial constraints, the Government should develop financial instruments ensuring sufficient co-financing of EU-funded activities. This includes blending different financing sources that can improve absorption capacity and speed up climate action.

By providing evidence-based policy advice and monitoring implementation, the EU can facilitate policy creation and implementation. This is necessary to support the transformation of society and provide for a tangible new narrative for European integration and cooperation.

Ana-Maria Boromisa is research advisor and Head of the Department for International Economic and Political Relations in IRMO, Zagreb. IRMO is a public research institute dealing with scientific and professional interpretation as well as evaluation of contemporary international relations which affect various human activities and related developmental trends that are specifically important for the Republic of Croatia.

IRMO is a public, non-profit, scientific research organisation located in Zagreb, with its main work organised through four departments. IRMO provides strategic support to decision makers and ensures the dissemination of its research results. IRMO is also a member of TEPSA.

Cyprus: Global Energy Crisis Is an Opportunity to Tackle Climate Change

Giorgos Kentas

Russia's invasion of Ukraine has given rise to many challenges across EU Member States, with the resultant new emerging security structure reshaping priorities in both foreign and public policies. The rise of energy prices and concerns over food security seems to have caused cost-push inflation that in turn questions macroeconomic stability, public finances and social cohesion. This spike is worrying for EU capitals and creates feelings of a vicious downward spiral where supplies of products and services may not be sufficient to keep up with demand. Nicosia needs to look for solutions. However, this difficult situation could be turned into an opportunity for reducing oil dependency, speeding up offshore energy programmes and accelerating efforts to tackle challenges emanating from climate change.

Climate Change: Concerns and Expectations

The 2021 Eurobarometer survey reveals that Cypriots consider unemployment, migration and climate change the top three challenges for the EU, with climate change being considered the main global challenge for the EU's future. The European Investment Bank climate survey (2021–2022) provides some more key perceptions. Cypriots think that climate policies will improve their lives (75%), green transition will be a source of economic growth (73%) and policies to tackle climate change will create more jobs (66%). They would also welcome tax imposed on products and services that contribute to global warming (75%).

Cyprus' energy mix (2020) depends heavily on fossil fuels (90.64%) with renewable energy being a rather marginal 4.84% (windfarms) and 4.51% (solar systems—photovoltaics and solar farms). In the last 32 years, CO_2 emissions have

G. Kentas (✉)
University of Nicosia - Cyprus Centre for International Affairs, Nicosia, Cyprus
e-mail: kentas.g@unic.ac.cy

increased by 63.27%. All these are reflected in a steady electricity bill hike and high emission fines. The Cyprus Electricity Authority must pay EUR 183 million for greenhouse gas emission rights in 2022, which is 150% more than the 2021 penalty. Electricity production in 2022 will produce some 2.96 million tons of carbon dioxide. Between 2021 and 2022, electricity bills increased by 48%. Given the present situation, Cyprus could not meet commitments made to the EU for increasing energy efficiency by 32.5% and producing a 55% (minimum) reduction in emissions by 2030, based on the 1990 levels.

A 2021 Cyprus Institute study makes some shocking predictions about the impact of climate change. In the next 30 years, the island's mean temperature is expected to rise by 1 °C to 3 °C. After 2050 the increase could be 3 °C to 5 °C and by the end of the century, this may even reach 7 °C. Undoubtedly, there will be some serious implications for peoples' lives, such as low precipitation and severe drought, which will ultimately lead to desertification. From 2020–2050 rainfall could decrease between 10–15% and by 2050, fertile land may be reduced by 72%, further decreasing to 82% by the end of the century. Water stress and desertification will reduce food production, bring heat waves and change the life in cities and rural areas.

Green Deal and Green Transition

Cyprus introduced its own roadmap for meeting the eight policy initiatives that comprise the EU's Green Deal Goals. In November 2020, the government presented a 'National Governance System for the European Green Deal in Cyprus 2021–2030' (NGS), which was approved by the Union. Key policy initiatives from successful implementation of the national system are aimed at: reducing greenhouse gas emissions, increasing the share of renewable energy sources for energy consumption, and improving energy efficiency. So far, progress is very slow in all three priority areas, with a major setback being the failure to import and use natural gas in power generation systems. For the foreseeable future, Cyprus' three thermal power stations will continue to use oil and gasoline. According to NGS calculations, electricity demand in Cyprus will double between 2021–2050 (from 4000 million KWh to 8000 million KWh).

However, there is a more promising dimension for addressing climate change challenges through the Cyprus-EU Partnership Agreement (2022) and the Just Transition Fund for Cyprus (2022), which are part of the Union's Cohesion Policy and include: the European Regional Development Fund; the European Social Fund Plus; and the Cohesion Fund. These arrangements provide for an investment of over EUR 387 million in energy efficiency, renewables and the reduction of carbon emissions. The focal point of that investment is creation of a circular and sustainable economy that will preserve local biodiversity and ultimately contribute to policies of adaptation to climate change.

NGS 2030–2050 targets are supported by the Just Transition Funds (EUR 101 million), which inter alia will be invested in renewable energy projects and storage technologies. There are also provisions that support small and medium-sized enterprises' efforts to increase their use of renewable energy technologies. An innovative element relates to the establishment of a Green Technical School, which will create educational and training programmes for empowering young people with new skills in green technologies.

A Climate Change Initiative for the Eastern Mediterranean and the Middle East

The battle for taming climate change's impact relates to history, culture, geography, economics and politics. In 2019, the government of Cyprus undertook a 'Climate Change Initiative for Eastern Mediterranean and Middle East (EMME)', an area which the World Meteorological Organisation considers a 'climate hot-spot with particularly high vulnerability to climate change impacts'. EMME comprises Bahrain, Cyprus, Egypt, Greece, Iran, Iraq, Israel, Jordan, Kuwait, Lebanon, Oman, Palestine, Qatar, Saudi Arabia, Syria, Turkey and UAE. The aim of this Initiative is to develop a Regional Action Plan for EMME, which will include concrete policy initiatives and actions that target challenges relating to climate change.

In the context of this Initiative, 13 task forces were created that involve more than 220 experts from EMME countries and some 20 more representatives of relevant international organisations. In the context of a Regional Climate Action Plan, these task forces have produced reports and made policy recommendations. The initiative is guided by Ministerial Meetings, where decisions are made for the next steps. To date, two Ministerial Meetings have been held, one in February 2022 and another in June 2022. This initiative is supported by the EU, the United States, the United Kingdom and Australia, who are represented in the Ministerial Meetings. In total, the initiative is supported by 18 countries. In the latest meeting, participants decided to introduce a 10-year plan, to be adopted at the next meeting in October 2022.

Recommendations

Cyprus has a clear and concrete programme for reaching climate change targets. In addition, the government of Cyprus has undertaken an ambitious initiative that epitomises a regional model for addressing climate change in EMME. Political commitment and consistency are the key elements of success. What matters most of all is to see results that will reverse the worrisome prediction of the Cyprus Institute study on climate change.

Nicosia needs to bridge a huge gap between pledges and outcomes. To achieve better results towards meeting the 2030 and 2050 targets, Cyprus' government must improve public sector efficiency and work closely with the private sector in improving and simplifying not only the process, but also procedures for creating renewable energy projects. Green transition has the potential to engineer economic growth and create job opportunities, but only if the government is prepared to adopt a strategic approach with specific benchmarks and performance indicators. Finally, Cyprus must urgently accelerate its offshore energy programme, working closely with licensed companies, the EU, Greece, Israel, Egypt and other EMME countries which have partnered Europe.

Giorgos Kentas is an Associate Professor in International Politics and Governance at the Department of Politics and Governance at the University of Nicosia. He is a Director of a Master Programme in Public Administration. His research focuses on strategic management, politics and governance at national and European levels. He follows EU developments and studies their implications for Member States and world politics. He has recently published papers on Brexit and its implications for Cyprus and strategic planning in the public sector of Cyprus.

The University of Nicosia is the largest private university in Cyprus. It offers more than 100 conventional and distance learning online programmes at Bachelor, Master and Doctorate levels. It hosts more than 11,500 students from all over the world and is also a member of TEPSA.

Czechia: Saving the Climate or the Czech Industry?

Zdeněk Sychra and Petr Kratochvíl

The Czech Republic, as with many other post-communist countries, has undergone a fundamental transformation in its relationship to the environment, moving from communist ignorance with its disastrous impact on nature, through the growing importance of environmental issues during the transition to democracy and eventually to the current perception of climate protection as a vitally important issue. Nonetheless, some members of the political elite, especially the older generation of politicians, both left and right, have yet to perceive the imperatives arising from climate change as forming a key priority, with some even arguing that the threat is exaggerated. This is matched by policy actions, which tend to be reactive, typically adopted in a reluctant manner under external pressure.

The Paradoxical Elites-Public Scepticism Gap

The Czech national debate is framed by a prevailing scepticism among political elites, especially the traditional political parties, towards more stringent climate protection measures. Public opinion in this area has been strongly influenced by the long-time Prime Minister and President Václav Klaus' denial of human influence on climate change. Moreover, the ideological narrative of both right-wing and left-wing governments has long emphasised the primacy of the economy over the environment. Correspondingly, environmental activism is often maligned, with the green movement pejoratively referred to as 'eco-terrorists'. Those most frequently denigrated are domestic non-governmental organisations such as 'Hnutí Duha' (Rainbow Movement), 'Děti země' (The Children of the Earth) and a number of single-issue grassroots organisations. The shift towards accepting carbon neutrality

Z. Sychra (✉) · P. Kratochvíl
Institute of International Relations Prague (IIR), Prague, Czechia
e-mail: sychra@kap.zcu.cz; kratochvil@iir.cz

as a more pressing topic remains rather slow. Significantly, in 2021 fossil fuels still constituted more than half of the country's energy production (54%, from nuclear energy at 40% and from renewables at 14%). Former Czech Prime Minister Andrej Babiš said as recently as December 2019 that he did not foresee carbon neutrality being achieved by 2050 and that the Czech Republic would more likely have reduced emissions only by 80% at that point. While the current centre-right governing coalition's stance is somewhat more forthcoming and programmatically supports the economy's green transformation, its representatives do not approach these issues as being of strategic importance either. Indicative of the government's stance, Conservative Prime Minister Petr Fiala, when asked whether the climate change was man-made, replied that 'the answer is not entirely clear'. The relatively fragile position of this environmental agenda is also reflected in the fact that the Ministry of the Environment has never been perceived as an important portfolio by the relevant political parties (with the historical exception of the brief government tenure of the Green Party).

The Czech public, though, considers climate change to be a more serious problem than the political elites. The recent serious summer droughts have become a significant turning point for the general perception of climate change in Czech society, having been clearly visible and affecting all segments of Czech society. According to the representative study 'Czech Climate 2021' 88% of citizens perceived there to be ongoing climate change, 21% saw themselves as favouring stronger action than 2 years ago and 70% supported taking early action even for the problems that will probably arise only around 2050. However, there is a minority position among the public—strongly supported by some politicians—which associates tackling the climate crisis with negative socio-economic consequences and negative impacts on living standards.

Clearly, Russia's war against Ukraine has changed the context in a fundamental way. The move away from energy dependence on Russia triggered action on many fronts: increasing capacity of oil and gas pipelines from EU countries (expansion of the TAL pipeline, construction of the Stork two pipeline from Poland); leasing floating liquefied natural gas terminals; supporting the construction of emission-free energy sources (mainly nuclear and photovoltaics); investing in energy innovation and savings; stopping the purchase of Russian fuel for the Temelín nuclear power plant; and exercising greater control over the energy sector (state entry into key power plants and takeover of the Russian share in Škoda JS, a key company for the Czech nuclear sector, by the ČEZ group).

Czechia's Environmental Reluctance and the EU as a Pace Setter

Ideological scepticism and a lax approach to climate protection have concrete consequences for the country's position compared to other European countries. The Czech Republic is regularly below the EU average in Climate Change Performance Index ranking, which analyses national measures to combat climate change. Indeed, together with Hungary and Poland, it was amongst the worst EU Member

States in 2021. The country is also among the largest emitters of greenhouse gases in the EU and has the greatest number of locations with the most polluted air after Poland. This clearly stems from the Czech Republic also being heavily industrialised, which results in high energy consumption, with some sectors such as the automotive industry being considered as objects of national pride. In addition to the industrial structure, another environmental problem is the Czech Republic's overproduction of energy, inherited from its communist past, which also contributes to the high per capita emissions. The energy sector (burning lignite in power plants, 39.5%) accounts for the largest share of greenhouse gas emissions, followed by transport (15.7%) and industrial production (12.6%, data available from 2018). Not surprisingly, the Czech Republic is one of the largest European exporters of energy.

The last long-term problem is the absence of strategic plans to promote low-emission or emission-free technologies. For example, a national hydrogen strategy, which could support its development in transport or energy, was introduced by the government only in 2021. Although the exit from coal mining has been set by the government for 2033, there is no concrete plan to achieve this aim. Furthermore, decarbonisation measures in transport and support for electromobility are weak. The reputation of renewables (especially photovoltaics) in the Czech Republic has been fundamentally damaged by the case of the so-called solar barons who profited from state energy subsidies around 2010. The state-guaranteed, but completely unregulated purchase of electricity at a distorted price damaged trust towards almost all renewable sources. It was only Russia's aggression against Ukraine and a reaction to the steep rise in energy prices that significantly kick-started their boom again.

European climate targets are often perceived in Czech society as an extreme, which does not reflect reality and hence hampers the country's economic development, especially for a country with significant industrial, energy-intensive production. The Green Deal is often framed as an environmental dictate from the EU and a threat to the Czech economy rather than an opportunity for a fundamental economic and environmental transformation. While the need for transition to a climate-neutral economy is already taken as a given that has to be accepted, reservations continue to be aired in the public arena. Within the National Recovery Plan, 44.5% of allocated funds are earmarked for green investments. This includes mainly projects to support energy renovation of buildings, sustainability of agricultural and forestry landscapes (biodiversity, water retention), reduction of energy consumption, transition to cleaner energy sources, building infrastructure for cleaner transport mobility or combating drought. At the same time, the Czech government is not willing to put up a fight and thus takes the existence of EU environmental commitments pragmatically as reality to which it must adapt. Conversely, it expects from the EU some acknowledgement of this specific position and significant financial assistance in the transition to a green economy. At the same time, the Czech Republic rejects some of the legislative proposals in the framework of Fit for 55, such as restructuring the EU minimum tax rate on energy products linked to the needs of EU climate policy. A key Czech effort has been the recognition of nuclear energy as an emission-free and stable source: the Czech Republic has two nuclear power plants that produce a third of its electricity and the country intends to strengthen the position of nuclear energy even further. Nonetheless, signing up to the EU's climate neutrality targets is gradually, albeit very slowly, being acknowledged as an opportunity for economic transformation. Yet, it is not reflected in concrete policy outcomes, concerning translation of strategic plans in individual areas, weak infrastructure and low support for electromobility, a number of exemptions from emission limits, erosive degradation of agricultural land as well as low efficiency in waste sorting and recycling.

Recommendations

The Czech government should respond to climate challenges in a preventive and strategic manner, not reactively and belatedly. In addition to the damaging environmental impacts, sluggish implementation may jeopardise Czech industry's competitiveness in the future. The change should be relatively easy as the general public is more concerned about the environment than those in political circles.

Transformation of the Czech economy in the context of tackling climate change must be as fair as possible in terms of all sectors and regions, some of which are heavily dependent on one type of industrial production. Otherwise, socio-economic impacts will lead towards greater support for populist and other anti-systemic political formations.

The government should therefore introduce a much more intensive form of communication towards Czech people, who are concerned about climate change, but yet find the issue's complexity difficult to grasp. Specifically, not only the risks, but also the opportunities this change brings should be addressed in order to counter the negative framing of this debate with a positive message.

Zdeněk Sychra is a member of the Department of Political Science and International Relations, University of West Bohemia in Pilsen. His academic interests include the issues of European politics, the Economic and Monetary Union and political governance in the EU. As an author and co-author, he has published numerous articles and book chapters on European Union politics.

The University of West Bohemia is one of the most visible universities in the Czech Republic. The Department of Politics and International Relations of the Faculty of Arts is an academic institution offering a broad range of undergraduate, graduate and post-graduate study programmes, conducting research in political science, international relations and territorial studies.

Petr Kratochvíl is a Senior Researcher at the Institute of International Relations Prague (IIR) and a Member of the Trans European Policy Studies Association's Board. He has written dozens of monographs, book chapters and journal articles. His research interests cover theories of international relations, European studies and the religion-politics nexus.

The IIR is an independent public research institution which has been conducting scholarly research in the area of international relations since 1957. Originally founded by the Ministry of Foreign Affairs of the Czech Republic, the IIR also provides policy analysis and recommendations. It aims to form links between the academic world and public/international political practice. IIR is also a member of TEPSA.

Denmark: Climate as a Given

Ditte Brasso Sørensen and Jens Mattias Clausen

In her celebratory speech following victory in the 2019 parliamentary election, the Prime Minister-designate Mette Frederiksen hailed this election as the first ever Danish 'Climate Election'. The label was apt. In the weeks before the vote, climate and environmental issues climbed to the very top of the political agenda and became the pivotal issue for voters across all age groups.

Climate Is Not Politically Contested

This perceived importance marked a significant change when compared with earlier elections. Surveys conducted in the run-up to the 2019 parliamentary election reported that 46% of the population identified climate and environmental issues as being among the three most important issues defining their vote (up from 24% in 2017). While other political concerns have since taken priority, such as managing the Covid-19 pandemic and reviewing defence policy given the war in Ukraine, these figures underscore the strong standing of climate and environmental issues amongst the Danish population.

Obviously, there are generational and political differences underlying the perceived importance of climate and environmental issues. Younger sections of society are more engaged in climate issues, as are people who vote for more left-wing parties. However, the differences are relatively small and do not undermine the general trend of climate concern among the Danish electorate. Indeed, according to the 2020 Climate Barometer, published by the Danish think tank CONCITO, 86% of the 60+ segment deem global climate change to be either a 'very' or 'somewhat' serious problem compared to 90% of the youngest segment (18–29). Differences are

D. B. Sørensen (✉) · J. M. Clausen
Think Tank Europa & Concito, Copenhagen, Denmark
e-mail: ditte@thinkeuropa.dk; jmc@concito.dk

more prevalent when comparing political parties. Where 61–76% of far-right voters agree that climate and environmental issues constitute a serious problem, this is true for 95–100% of the red-green left-wing voters. The need to address climate and environmental issues is significantly less politically contested than in many other European states. Furthermore, a trade-off between climate action and welfare is not generally recognised by the Danes. To the contrary, 68% of Danes understand the green transition rather as a driver of growth and welfare. Coupled with a widespread sense of urgency, this makes political turbulence with regard to Danish climate ambition unlikely, at least wits regards to headline targets.

From Global Front Runner to European Partner?

Denmark has a long-established history of progressive climate policy as well as an advanced and globally competitive 'green-tech' industry. The combination of political ambition and technological know-how has made Denmark a both real and self-proclaimed climate front runner, at least when it comes to domestic emissions.

Wind and solar (with the latter accounting for 3.6%) made up 47% of all electricity consumption in 2021, while the total share of all renewable energy was 43% of final energy consumption. However, it should be noted that solid biomass accounted for 23% of energy consumption, by far the largest renewable energy source, and biogas for 4%, more than doubling since 2018. To solidify this position, a broad parliamentary majority adopted the Danish Climate Act in 2020. This Act sets a 70% greenhouse gas emissions reduction target by 2030 (compared to 1990 levels) and guarantees climate neutrality by 2050. Recently broad political agreement was reached on the introduction of a national carbon tax, supplementing the EU Emission Trading System. In addition, Denmark has accelerated the phase-out of gas as well as the expansion of wind and solar energy leading roughly to its quintupling by 2030 compared to current capacity. It has proved hard, though, to bring change about within the agriculture and transport sectors: Agricultural production takes up more than 60% of the total land area of Denmark, while transport is the only major sector that has failed to reduce emissions since 1990. Transition in both sectors relies to a significant degree on EU policies.

While Denmark has a track record of supporting European climate policies, the EU has not been showcased domestically as an enabler of the country's approach. However, of late RePowerEU together with the European Recovery Plan, which has helped finance flagship projects such as a green tax reform, have highlighted the EU's importance as a central facilitator of Danish climate policy. The decision to host the 2022 North Sea Summit in Esbjerg is testament to a growing political recognition that the European arena is a powerful facilitator of green ambition and that concerted action at EU level is key to achieving national, regional and global climate targets. This Summit brought together Heads of Government from Germany, Belgium, the Netherlands and Denmark, as well as Commission President Ursula von der Leyen.

Frugality Poses a Risk to the Danish Climate Ambition

Denmark will take on the EU presidency at a crucial point in time for the second half of 2025. While it will hopefully be in the aftermath of the war in Ukraine, the effects of this war, such as an alarming food crisis together with high and volatile energy prices, will likely still be felt. Furthermore, Denmark's presidency will follow elections to the European Parliament in 2024 and inauguration of a new Commission along with the run-up to a pivotal United Nations climate summit (COP30) at the end of 2025, where all parties are expected to submit new Nationally Determined Contributions. Climate will thus be at the very top of the European and global political agenda when Denmark takes up the reins.

Hence, this presidency presents a unique opportunity to promote European action on Danish climate strongholds, namely: wind energy; energy efficiency; biosolutions and food; carbon capture; as well as collective green infrastructure, such as district heating. Denmark's ability to utilise the European stage effectively in achieving its climate ambitions depends on a political willingness to rethink its frugal position. As part of the so-called 'Frugal Four' informal alliance (along with the Netherlands, Sweden and Austria), the country calls for more conservative EU fiscal policies. Keeping national contributions to the EU at current levels, as well as ensuring strict enforcement of the EU's fiscal rules continues to be a long-standing political priority for Denmark.

Danish frugality has also affected its position on EU climate policy. The Danish government has, for example been critical of the establishment of the Social Climate Fund, and it was reluctant to accept joint European debt issuance in relation to Next Generation EU and fought hard to limit any grants offered as part of the European recovery funds, which are in part designed to facilitate green transition in European member states. Denmark is reluctant to support European initiatives that risk distorting competition (such as industrial policies) or initiatives which require expansionary fiscal policies, even if these are supporting green transition. Indeed, reluctance to accept that green priorities will at times need to take precedence over frugal priorities could inhibit Denmark's chances of advancing its ambitious climate polices at European level.

Recommendations

To ensure that the EU becomes a 'Climate Union' Denmark should work to develop and promote climate alliances with progressive member states, allocating sufficient political capital to the project.

To sustain its position as a climate front-runner, Denmark should address the issue of consumption-based emissions, including the use of goal setting, in order to reflect the Danish carbon footprint adequately. In addition, taking into account continual sustainability issues with the use of biomass, Denmark should develop a plan to phase out its use with the aim of reaching 50% elimination by 2030.

Populations in frugal states are mostly concerned with issues such as corruption and misuse of EU funds. Denmark and its European allies should seek to re-brand their firm stance on fiscal policies to include a stringency on green conditionality and strict prioritisation in the use of EU climate funds.

When Denmark takes over the Council's rotating presidency for the second half of 2025, this will provide its government with ample opportunity to highlight the EU's role in domestic climate policy. Moreover, it will present openings for promoting Danish strengths in areas such as wind, energy efficiency and biosolutions (e.g. enzymes and alternative proteins). Denmark should use this upcoming presidency to advance the EU's work on areal-effective production methods and guard against deterioration of the EU's environmental regulation.

Ditte Brasso Sørensen is a Chief Analyst at Think Tank EUROPA. She holds a Ph.D. in Political Science from the University of Copenhagen and has previously worked as an EU advisor to global companies, research institutions and national, local and regional authorities.

Think Tank EUROPA is a Copenhagen-based independent think tank. Its purpose is to provide fact-based analysis on European politics and EU policy in an effort to facilitate debate.

Jens Mattias Clausen is EU Programme Director at CONCITO. He has a master's degree in political science from the University of Copenhagen and has previously worked as Chief Strategic Advisor to the Danish Minister of Climate, Energy and Utilities and before that Head of Climate Policy at Greenpeace Nordic and Head of the UNFCCC Delegation for Greenpeace International.

CONCITO is Denmark's green think tank. Founded in 2008, its purpose is to provide science and knowledge-based analyses and information on the most effective and cost-efficient transition towards a climate-safe society in Denmark and in other parts of the world.

Has Estonia Already Lost the Path to Achieving Its Climate Goals?

Viljar Veebel

Estonia's geographical location and closeness to the Baltic Sea create difficult and variable climactic conditions. Most of its territory experiences a continental climate, humid with cold winters, albeit Estonia's islands and west coast enjoy a more maritime climate with mild winters. Differences in climate conditions are aggravated by the Baltic Sea that warms up Estonia's coastline areas during the winter period but cools them down in spring and summertime. Since 1950, air temperature as the basic climactic indicator has been steadily increasing in Estonia, with a mean reading over the second half of the century up by 1–1.7 °C at different stations around the country. Global air temperature over the same period was up by about 0.4 °C, according to climate experts Jaagus and Roper. Rainfall during this time also increased, with on average there being less days with snow cover each year. Winter storms have become more common, spring starts earlier, autumn arrives later, whilst summers are longer, drier and warmer than in the past.

There are two principal options for coping with climate change, which are: either to mitigate the negative effects, namely greenhouse gas reduction; or to adapt accordingly. In 2017, the country made a commitment at national level to reduce greenhouse gas emissions by 70% as its 2030 target and 80% by 2050, compared with the 1990 levels.

According to government agency Statistics Estonia's 2021 publication, the two sectors that produce most CO_2 emissions are the energy sector (electricity, gas, steam and air conditioning supplies) and the manufacturing sector, producing 59% and 18% of total CO_2 emissions (based on 2019 data), respectively. This strategy document lists more than 20 measures on how to achieve climate targets, including implementation of technologies with low CO_2 emissions, gradually more extensive exploitation of domestic renewable energy resources, enhancement of bioenergy production and reduction of waste generation. Furthermore, in 2021 the national

V. Veebel (✉)
Estonian Foreign Policy Institute, Tallinn, Estonia
e-mail: Viljar.Veebel@ut.ee

© The Author(s), under exclusive license to Springer Nature Switzerland AG 2023
M. Kaeding et al. (eds.), *Climate Change and the Future of Europe*, The Future of Europe, https://doi.org/10.1007/978-3-031-23328-9_8

31

parliament approved a strategy document, 'Estonia 2035', declaring inter alia that by 2050 there will have been a transition to climate-neutral energy production in Estonia. As outlined in this publication, measures to achieving climate neutrality include reverse auctions for renewable energy; support for the production and consumption of biomethane; renovation of inefficient heat pipelines; together with support for wind and solar energy. Moreover, there are ongoing investigations into the possibility of introducing nuclear energy in Estonia to achieve the climate goals, with the respective working group due to publish a preliminary report sometime during 2022. The Ministry of Environment further indicated in its regular report that Estonia's greenhouse gas emissions in 2020 were 71% less than national net emissions in 1990, with the largest producers being the energy, transport and agricultural sectors.

As far as the climate-neutrality goal is concerned, the country's biggest practical concern relates to the local oil shale mining industry. Oil shale is the main mineral mined in Estonia and most of its resources are concentrated in the north-eastern county of Ida-Viru. The sector was first hit in 2019 when CO_2 prices peaked and producing energy from oil shale ceased to be cost-effective. As a result, hundreds of workers in Ida-Viru lost their jobs. In any event, oil shale mining is not particularly environmentally friendly and hence environmental interest groups in Estonia as well as EU institutions, largely the European Commission, have been putting pressure on Estonia to close down the industry.

Whilst this industry is not particularly relevant for Estonia in economic terms, as it produces only about 5% of the country's GDP and employs around 2.5% of the total labour force, it does have enormous importance in socio-economic and security-related terms. A 2020 Praxis study concluded that closing down the oil shale industry in Ida-Viru could place at least 8000 people in danger of poverty. Closing down local oil shale mining companies also produces wider negative impact on local government tax revenues in Ida-Viru that in turn affects the region's economic growth potential. Furthermore, the situation is made particularly complicated by the presence of mostly Russian-speaking residents, many of who work in the oil shale mining industry. Consequently, by closing down this industry, the government seriously risks worsening an already fragile security situation, caused by ethnic tensions due to Russia's war in Ukraine. In this light, it is obvious that Russian-speaking people in Estonia (around 25% of the population) will find no reason to support the country's climate goals.

Has Estonia Lost Its Focus on Achieving Climate Goals?

On the one hand, the Ministry of Environment's most recent report from 2021 declares that Estonia will be unable to achieve the goals set in the strategy documents by 2030 and 2050. Hence, additional measures are needed to help Estonia achieve these goals. On the other hand, it almost feels that due to the recent energy crisis (combined with the war in Ukraine) that people have lost their motivation to contribute to the country's green transition. For example, transition to renewable energy sources was considered to be very popular among most social and political groups in Estonia up to November 2021, because it was expected that it would contribute not only to a cleaner environment, but also increases in financial savings. However, since costs for heating and electricity more than tripled for Estonian households over the winter of 2021/2022 compared with the previous year, people have generally become very critical as regards the sustainability of green transition and its costs to households. As a result, public opinion is now divided between supporters and opponents of the EU's policy on renewable energy and as long as the energy bills are high in Estonia, the voices of those who oppose green transition will become much louder. Although the Next Generation EU provides additional resources and opens up some new possibilities in Estonia as regards green transition (such as, for example, support for the implementation of new, green technologies), its cumulative long-term effect is rather modest. 2020 estimates by Niestadt show that Next Generation EU could increase the country's gross national product over the next 10 years by about 0.8–1.3%.

In this respect, the fact that most local politicians here support the EU's climate goals as well as the principle of Estonia's climate neutrality in the future and are ready to push forward with these goals, there is a real risk that tensions could boil over in Estonian society. Moreover, these tensions will undoubtedly be exploited by those political parties (such as the Conservative People's Party of Estonia, EKRE) for the sake of their own political self-interests in order to gain more public support.

Recommendations

Both at EU and national levels, in the process of green transition, focus on a variety of economic sectors, not just on a limited number of 'easy' options such as information technology, so as to diversify the risks.

Both at EU and national levels, contribute more to innovative technologies and research and development projects, as well as enhancing research and development practical cooperation between universities/research institutions, state and enterprises.

At national level, there is a need to be more flexible and adjust to respective circumstances, such as risks related to energy dependence and high inflation. Provide well-targeted measures to avoid increases in poverty throughout Estonia.

Viljar Veebel is a researcher at the Department of Political and Strategic Studies of the Baltic Defence College and a National Researcher for the European Council on Foreign Relations (ECFR). He has worked as an Academic Advisor for the Estonian government in the European Future Convention and as a researcher for OSCE, SIDA, Estonian Foreign Policy Institute and Eurasia Group. His main research interests include European security and defence initiatives, use of economic sanctions as a foreign policy tool, EU-Russia relations and related sanctions. In the current volume, he represents the Estonian Foreign Policy Institute.

The Estonian Foreign Policy Institute (EVI) is an independent, non-profit foundation that research focuses on the interests of regional security, European Union integration and enlargement, as well as developments in Russia. EVI is a member of TEPSA.

...

Finland's Fight Against Climate Change: Ambitious Yet Pragmatic Approach

Juha Jokela

Successive Finnish governments have seen the fight against climate change as a critical factor affecting not just the future of Finland, but also the whole of Europe and the rest of the world. Climate change became one of the key topics during the latest parliamentary elections in 2019 and the current coalition government, led by the Social Democratic Party and including the Green Party, has set an ambitious goal for Finland to become carbon neutral by 2035. This domestic objective is now enshrined in national legislation. Moreover, in general terms forging a green economy together with investments, innovations and solutions linked to clean technologies are seen as creating a source of economic growth and prosperity for Finland.

By contrast, the need for secure energy sources especially during the cold, dark winter months has highlighted the need for retaining nuclear power along with investment in renewables. The government has also opposed EU proposals which could limit the use of Finland's vast managed, commercial forests. In doing so, it has aimed to clarify and justify the sustainability of its forestry regarding protecting biodiversity and enlarging carbon sinks. To secure the EU's major leap towards carbon neutrality, more information concerning the relative weaknesses and strengths of Member States is needed to increase understanding of how much change is needed together with the related costs and benefits. Helsinki should also initiate a process to evaluate the impact of increasing global and regional strategic competition *vis-à-vis* the EU's climate leadership. Finland's pathway to carbon neutrality will be determined largely by the successes and failures in ongoing economic, political and societal transformation.

On the positive side, the Finnish economy and key industries, including the energy sector, are ahead of the curve regarding European and global transformation.

J. Jokela (✉)
Finnish Institute of International Affairs, Helsinki, Finland
e-mail: Juha.Jokela@fiia.fi

© The Author(s), under exclusive license to Springer Nature Switzerland AG 2023   35
M. Kaeding et al. (eds.), *Climate Change and the Future of Europe*, The Future of Europe, https://doi.org/10.1007/978-3-031-23328-9_9

The government is also aiming to steer this transition, to the extent that it has initiated a process in which sectoral low-carbon roadmaps are being developed in cooperation with companies and organisations within the relevant sectors. Against this backdrop, key industries' strategies are increasingly being built upon growth and profit trajectories stemming from carbon-neutral products, services and innovations. Finland's iron and steel manufacturing, valued at EUR 6.2 billion and ranked ninth largest in Europe, for instance, is now gearing towards producing carbon-neutral products for the emerging and growing green markets. Innovations related to clean technology are in general seen as providing competitive advantage for the Finnish economy. Moreover, the energy sector has accelerated its de-carbonisation. For example, Helsinki City's energy company providing central heating and electricity for the capital region has significantly accelerated the closure of its two-coal operated powerplants with the result that it is now aiming to be fully carbon neutral in 8 years. The power plants will be replaced by a new decentralised system providing heat from various carbon-neutral and renewable sources. Digitalisation and data are utilised to optimise productivity and efficiency.

While the turn toward renewables plays a major role in Finland's aspirations to be energy independent in a matter of years rather than decades, the country has also decided to retain substantial nuclear energy production capacity. Even if the joint project with Russia's Rosatom for the sixth reactor has now effectively been cancelled, the long overdue fifth reactor build by French Areva and German Siemens should be fully operational in 2023 after extensive delays. Moreover, state-owned energy company Fortum has recently filed an application for extension of licences for the countries' two oldest reactors. The rationale behind these decisions is related to security of energy supplies during cold winter months and an aspiration to become energy independent. The importance of both objectives has been highlighted by the rapid deterioration in EU-Russia energy relations. Finland has been largely able to replace Russian fossil fuels (coal, oil and gas). Gas represents around 5% of Finland's energy mix, and it is now provided through the recently build 'Balticconnector' pipeline connecting Finnish and Estonian gas networks as well as via LNG terminals. Russian pipeline supplies to Finland were cut when Finland refused to yield to Russian demands for payments in roubles. Russia has also stopped providing electricity to Finland, which might create some short-term challenges despite the well-functioning joint Nordic (and Baltic) electricity markets. Yet investment in wind power, renewables and reliance on nuclear power should make Finland largely electricity independent in couple of years.

Given the country's large size and long distances between population centres, de-carbonising transportation is seen as problematic especially due to cold weather conditions limiting the range of current electric cars and heavy traffic. Given the loosely populated, vast territory—almost the size of Germany, but with only some 5.5 million inhabitants—electrified rail connections provide only a partial solution.

Finnish forestry and agriculture play a key role in the country's climate neutrality targets, as well as sustainability targets related to biodiversity loss. Taxonomies, strategies and legislation related to the EU's Green Deal, have resulted in some political cleavages in Finland, though. The government has defended the country's land and forest usage policies as being both sustainable and helpful for the EU's carbon neutrality targets. However, researchers and civil society organisations have questioned the ecological sustainability of current national land and forest use policies. Furthermore, the Commission has put more weight on biodiversity loss, urging limitations and changes in current practices. Finland plays host to many major forest industry companies. In addition to paper and pulp, the industry increasingly provides products based on renewable wood materials, which can be used in plastics, composites, liquid fuels, nanocelluloses and lignin, along with medicines and textile fibres.

At political and societal levels, an overall perception of the need to fight against climate change has expanded during the past years, but with the electorate calling for pragmatic actions. There is also some evidence of increasing politicisation of climate and environmental matters in Finland.

The far right and populist Finns party has questioned the thinking behind Finland's ambitious climate policies. Prior to the last parliamentary election in 2019, they were the only party which did not join a joint national action plan on climate change signed by all other eight parties in the parliament. The Finns Party has suggested that more attention should be directed to the key global and European greenhouse gas emitters than already relatively green smaller economies, such as Finland. They have also tried to frame supporters of climate policies as followers of 'green ideology' and/or privileged economic and political elites, accusing them of 'climate hysteria' and not defending Finland's economic interests hard enough.

Conversely, there has been increased visibility for numerous civil society groups from major organisations such as the Finnish Association for Nature Conservation, Green Peace, and Amnesty International Finland. Movements such as Extinction Rebellion Finland utilising non-violent civil disobedience in their demonstrations highlight more ambitious action to address climate crisis and biodiversity loss. Political polarisation of climate matters has been observed in recent opinion polls examining the values and attitudes of Finns. Nevertheless, in 2021, 71% of Finns considered climate change to be the number one threat to our environment, which requires urgent action from all states.

At the same time, some dissatisfaction has emerged in Finland regarding the EU's proposal to establish a new Social Climate Fund, financed by the proposed Carbon Border Adjustment Mechanism and aimed at helping the most disadvantaged EU citizens during the transition toward carbon neutrality. Given that payments currently being channelled via the EU's recovery and resilience facility represent a significant financial transfer to the EU's weaker economies' green, social and inclusive growth, further transfers via a new fund could turn out to be politically difficult for Finland. Moreover, the government has reminded the European Commission that Carbon Border Adjustment Mechanism's income should be used to pay back the bonds issued by the Commission to finance the recovery and resilience facility.

In any event, the current Finland is clearly aiming to be on the right side of history regarding the fight against climate change. There are increasing expectations that the

EU will push through the ambitious Green Deal and lead the way to a global carbon-neutral future.

Recommendations

In order to enhance the transition to carbon neutrality as a tangible new narrative for European integration, the Finnish government and the EU should work harder to provide easily accessible information about (and communication on) the costs and benefits of transition for each Member State. This would provide EU citizens with an opportunity to assess the scope of transition and the levels of investment needed in their own countries. Assessments on the relative strengths and potential competitive advantages of Member States would also be helpful, as would the costs of inaction.

Russia's invasion of Ukraine has highlighted the need for urgent work toward sustainable energy policies at national and European levels, also with a view to security of supply. The implications of war have rightly resulted in demands to accelerate Finland's and the EU's transition towards non-fossil and carbon-neutral sources to end energy dependencies on Russia.

From Helsinki's perspective, the Union's key challenge is implementing the Green Deal in an environment marked by increasing strategic competition in the world's economy and politics. Finland should continue its strong support for the EU's global climate leadership. Thorough assessments are needed concerning the potential negative impact of increasing great power competition on multilateral efforts to tackle climate change. To this end, Finland should promote a new EU Global Strategy to fight climate change in an era of increasing strategic competition.

Juha Jokela is a Director of the EU research programme at the Finnish Institute of International Affairs (FIIA). He was a board member of the Trans European Policy Studies Association from 2014–22. His research interests include EU's foreign policy, differentiated integration, Finland's EU policy, and the political implications of Brexit.

FIIA is a research institute whose mission is to produce high-quality, topical information on international relations and the EU. FIIA is part of TEPSA.

France: Not Living up to Its Ambition

Charlotte Halpern

It is common knowledge that climate change policies in France and environmental protection more generally have long held an ambivalent position in comparison with other European countries. This is mainly due to a discrepancy between the 'heroic rhetoric' that characterises political discourses about environmental protection and the ability to make these political goals operational.

In this regard, President Macron's first mandate did not break with tradition. In 2017, a few days after the United States had decided to withdraw from the Paris Agreement, the newly elected president forcefully addressed the American people in English, urging them to 'make our planet great again' and with this, sought to establish France's climate leadership globally. Further to this initial statement, his approach was later defined as: 'Ecology, fight of the century. [. . .] In France, an in-depth transformation of our model [. . .] At the global scale, ensure the mobilisation of all'. By opting for the notion of 'ecology', he marked his ambition not to limit any action to climate issues, but also to address biodiversity and the environment at large. Five years later, following a radical change in geopolitical context and having faced massive protests in France, President Macron is far from seeing the country as living up to its international ambitions.

In France, a Series of Small Steps Is Not Adding up to 'An In-Depth Transformation' of the Model

Following the 2018 failed carbon tax project and the trauma represented by the 'Yellow vests' protests, ruling elites in France have been struggling to come up with a socially just approach to climate transition without undertaking a genuine in-depth

C. Halpern (✉)
Science Po, Centre for European studies and comparative politics, CNRS, Paris, France
e-mail: charlotte.halpern@sciencespo.fr

transformation of the existing system, including its institutional, political and economic components. Does this mean that nothing has been done? Certainly not. France's path towards climate neutrality now consists of two main strategies which combine to provide firstly a legislative framework for the achievement of carbon neutrality by 2050 and secondly a reduction of its carbon footprint. The National Low-carbon strategy, introduced in 2015 and revised in 2020, has identified the main drivers and levers for this climate change mitigation policy in regard to transport, housing and industry. Progress has been made to enhance and expand the climate and resilience policy framework, covering new issues such as soil protection and setting new targets for a circular economy and anti-waste, as well as enhanced food quality for schools catering services. A number of controversial projects were abandoned, including shale gas research and exploitation, the Notre dame des Landes airport project, the Montagne d'or project in French Guiana, as well as the EuropaCity retail project in the north of Paris. New administrative structures have been created, such as the French Biodiversity Agency. Monitoring tools, such as environmental budgetary reviews and impact assessments, have been introduced. Flagship initiatives in governance, such as the Citizens' Convention for Climate or the Higher Climate Council, were introduced as part of efforts to reach out to citizens and scientific knowledge so as to overcome well-known barriers and resistance to climate action.

The addition of new layers—structures, laws, targets—did not, though, automatically result in transformative change. Rather, it contributed to exacerbating fragmentation of the climate governance system without providing it with sufficient clout to overcome institutional resistance, bureaucratic routines and corporatist arrangements. By framing their demands in terms of 'feasibility', 'reason' and 'pragmatism', professional organisations and interest groups—in the agriculture, transport and energy sectors more specifically—successfully obtained the postponement of carbon emission reduction targets, as well initial proposals to ban harmful products and activities. Here are a few examples.

Less than a third of the measures resulting from the Citizens' Convention for Climate in 2020 were included in proposed bills, including those to ban short-haul flights when a rapid train alternative exists. The proposed ban on glyphosate was eventually suspended in 2019 pending a European decision. In the case of glue traps hunting, for which a ban was proposed in 2020 to bring France into compliance with the EU Birds Directive, Minister Pompili was violently accused by the powerful hunters' lobby of being dogmatic during a 2020 interview with RTL, whilst regional hunters' associations charged her with acting 'in the name of ideological motivations having nothing to do with the protection of animal species. She received little support from within the government and political elites more generally, with a number of prominent politicians and President Macron himself wanting to protect traditional forms of hunting as part of the country's heritage and art of living (discourse held at the 2017 Congress of the National Federation of Hunters). Following months of procrastination, the State Council eventually declared the ban in 2021, following a legal action initiated 5 years earlier by the Birds protection league.

Energy transition provides another example of procrastination and breaking with the commitments made at EU level. France still is still failing to achieve EU targets for developing renewable energy and as yet no clear decision has been made to close its coal mines. In the case of nuclear power and plans to reduce its share in the energy mix from 75% to 50% by 2025, in late 2020 President Macron sanctioned an extension of the deadline to 2035. One year later, he announced a major investment programme in the nuclear industry to develop a new generation of power plants. Both within and outside the state apparatus, pro-nuclear discourses have gained a new momentum following Russia's invasion of Ukraine, framing it as a major asset to limit the country's energy dependence on imported fossil fuel-based sources.

Lost in the Politics of Small Steps

In spite of ambitious political discourse, the fight against climate change is being slowed down by the politics of small steps, which have combined to prevent the achievement of objectives initially set. Such ambivalence also is also reflected in proposals stemming from political parties and civil society. It is absolutely certain that environmental issues are now unanimously recognised as a major socio-political issue and furthermore benefit as such from increased coverage, including non-specialist organisations. Concrete proposals for a significant reform of the climate governance system were laid out and extensively discussed during the 2022 campaign, among which was the need for ecological planning as a way to align strategic milestones across sectors and levels of governance with long-term objectives. This accounts for the fight against climate change being a major dividing factor among and between organisations when it comes to the nature, width and depth of transition, as well as its drivers—technology, prices, regulations, etc.—which need to be selected, in spite of unprecedented opportunities offered by a political system in the midst of restructuring. This is also reflected in public opinion. According to regular measurements and comparative survey results (ADEME, special Eurobarometers, etc.), environmental issues moved to the top of French people's priorities in 2019, with results in the European, municipal and, to a lesser extent, legislative elections bearing witness to this. According to responses from a two-wave IFOP survey in 2019 and 2020, respondents now primarily identify citizens and subnational levels of government as the main actors who are capable of acting in favour of the environment, while they consider the government's actions for the most part to be ineffective. In regard to President Macron's commitment and competence, they give him a rating of four to five points out of ten.

In this context, who are climate champions? Supporters of the climate movement and the green electorate more generally still tend to be younger, rather urban, mainly women and with a higher level of qualifications. Drawing on powers gained after the 2015 decentralisation reforms as well as the renewal of municipal and regional political leadership in 2020 and 2021, respectively, a growing number of subnational authorities have emerged as champions in the fight against climate change. Through their climate agendas, cities such as Nantes, Poitiers, Lyon and Grenoble, or regions such as Grand Est, Nouvelle Aquitaine and Val de Loire, are increasingly challenging the state in its ability to develop a holistic and country-wide transition strategy.

Within the climate movement itself, similarly to developments in other European countries, action repertoires are being reconfigured, having been motivated by the arrival of new associations such as Alternatiba and Extinction Rebellion, as well as the creation of joint platforms in conjunction with anti-poverty organisations such as Oxfam. While the former push for the use of radical and targeted actions, the latter have opted for a socio-ecological transformative agenda and as such have successfully sued the state for climate inaction (Affaire du siècle). Interestingly, a similar strategy was adopted by local authorities in calling on EU institutions to sanction France's lack of compliance with EU air quality requirements. By contrast, advocacy work outside the government focused on institutional reforms (Terra Nova), a strategy respectful of planetary boundaries and grounded in nature-based solutions (World Wide Fund for Nature France, IDDRI). Within the state apparatus, a set of four transition scenarios have been developed by ADEME to support a national strategy based on inclusiveness and sobriety. France Stratégie's report on sustainability proposes a profound restructuring of the climate governance system to enhance policy coherence and achieve transformative change.

At European Level, Neither a Pioneer Nor a Climate Leader

Both the United Nations Climate Change Conference (COP 26) in Glasgow and the French EU presidency could have provided major opportunities for France and President Macron to strengthen the country's green leadership. To this end, the Higher Council Climate's 2020 and 2021 reports urged for a more ambitious agenda to be adopted, should the country intend to play a leading role globally and at EU level. The reports highlighted France's delay in achieving its climate objectives by 2030, pointing more specifically at insufficient efforts being made to reduce emissions in road transport and agriculture.

When it comes to climate change, and notwithstanding President Macron's clear commitment to European integration, the country's position, namely 'a sovereignty agenda' for France in Europe and Europe in the world, is primarily guided by the promotion of its strategic interests and marginally shaped by its climate ambitions or efforts at supporting enhanced integration in this area. Similarly, to former ruling majorities, successive governments under Macron's Presidency have chosen to shift the burden of introducing unpopular measures, regarding, for example agriculture transition and air quality, to European institutions, which then provides an opportunity to shift blame and delay implementation, as they have done in the past. President Macron did, though, choose to prioritise adoption of a carbon border tax as part of his presidency and considering Russia's invasion of Ukraine, to work towards increased cooperation in reducing energy dependence. In the meantime, France has aggressively pushed for nuclear energy to be labelled as a green investment under EU taxonomy. Funds available as part of the EU Green Deal and under the climate bank feed into the country's objectives for decarbonising its industry, supporting the developing of new sectors (such as hydrogen) and exploring scope for innovative processes and technologies.

Recommendations

The following recommendations should be directed towards the French government and the EU.

Firstly, a clear political narrative about what climate adaptation means must be developed, linked to the widest possible debate on the difficult choices to come, including major constitutive reforms and the range of solutions available for the fairest possible transformation.

Secondly, the current climate governance system's inadequacy should be acknowledged, due to its high level of fragmentation, as well as its limited clout to transform existing corporatist arrangements and bureaucratic procedures effectively. This should be taken as a first step towards prioritising climate objectives at strategic and operational levels.

Finally, in addition to climate policies aimed at shifting or improving the current situation, existing policies must be reviewed according to their carbon impact in a first step towards a roadmap aimed at eliminating all those that do not, at the very least, aim at carbon neutrality.

Charlotte Halpern   is a tenured researcher in Political Science at the Centre for European Studies of Sciences Po and co-director of the Environmental policy research group at the Laboratory of interdisciplinary public policy evaluation (LIEPP). Her research focuses on State restructuring and comparative public policy processes in Europe, with a focus on climate and environmental issues. This work benefits from support provided by the ANR and the French government under the Investments for the Future programme LABEX (ANR-11-LABX-0091, ANR-11-IDEX-0005-02) and the Idex University Paris Cité (ANR-18-IDEX-0001).

Sciences Po is the leading French research university in Political Science, International Relations and Sociology. Since 2005, the Centre for European Studies and Comparative Politics aims at producing research on European issues, facilitating Sciences Po's insertion in European research networks and fostering debates on the future of Europe. It is also a member of TEPSA.

supporting the developing of new sectors (such as its droogen) and expiration, so up to innovative processes and technologies.

Recommendations

The Fight Against Climate Change in Germany: From Energiewende to Zeitenwende?

Funda Tekin and Christina Goßner

In Germany, two factors determine climate policy and the extent to which it features in political debate. On the one hand, party politics and their representation in government and parliament are relevant. Climate protection did not appear prominently on Germany's political agenda before the Greens entered the German Bundestag in 1983, driven by a strong anti-nuclear movement. The Greens are traditionally considered the 'mother party' of climate policy in Germany. On the other hand, German climate policy has a rather reactive element, because it is heavily influenced by disruptive incidents linked to the environment. The earthquake and nuclear fall-out in Fukushima in 2011 put an unexpected end to the tough debate on phasing out nuclear energy. Disputes over the infamous railway project Stuttgart 21 facilitated the Greens' success in regional elections for Baden-Württemberg in 2016. The anticipated deforestation of the Hambach Forest in the coal region of North Rhine-Westphalia caused massive protests against the coal industry in 2018. Finally, the movement Fridays for Future together with the flood catastrophe with more than 180 fatalities in western Germany over the summer of 2021 put climate protection prominently on the electoral programmes of all parties in the September 2021 federal elections. German citizens consider the climate crisis a matter of special urgency, which is reflected in the Greens' latest electoral successes both at federal and state level since 2021. According to surveys conducted in November 2021 on the need for action on climate protection, Germans would welcome an even more ambitious climate policy.

Germany's traditional self-perception as a pioneer in fighting climate change is not necessarily reflected in its performance. During the spring of 2021, the Federal Constitutional Court ruled the Federal Climate Change Act of 2019 insufficient to protect the freedoms of young people and future generations. The revised Act

F. Tekin (✉) · C. Goßner
IEP Berlin, Berlin, Germany
e-mail: funda.tekin@iep-berlin.de; christina.gossner@iep-berlin.de

requires Germany to bring about a reduction in greenhouse gas emissions of 65% by 2030 compared to 1990 and to achieve greenhouse gas neutrality by 2045. While the country's climate targets exceed those of the EU, Germany is a Member State with above-average emissions compared to its population. It alone is responsible for 2% of total global emissions, with the energy sector being the largest generator (30%). Emissions in other sectors such as industry (24%), transport (20%) and construction (16%) are also noteworthy. Understandably, Germany's climate policy is largely focussed on the 'Energiewende' (energy transition).

There are various reasons why Germany regularly misses its climate targets. Firstly, due to natural deposits, Germany has traditionally generated electricity through coal, with entire regions being dependent on this energy sector. Secondly, German climate policy is exposed to partisan battles, most prominently in relation to energy transition, which has led to an erosion of political support within core constituencies amongst coal regions, such as the Social Democratic Party in the Ruhr area. The former grand coalition of Social and Christian Democrats almost fell apart during the autumn of 2019 over discussions about a law on sector-specific emission reductions. Thirdly, Germany's federal system occasionally further complicates the matter. Not only does energy transition impose unequal costs on the federal states but also diverging climate programmes and laws at state, county and municipal levels often lead to conflicts of interest. This is particularly evident in the expansion of wind power, because state regulations in Bavaria and North Rhine-Westphalia, for example, set out stretching conditions for the construction of wind turbines. However, in North Rhine-Westphalia the fact that the Greens form part of the new government coalition with the Christian Democratic Union has already had an impact, with the expansion of wind power being regarded as a prestige project.

German climate strategy is also strongly influenced by the country's overall economic policy and certain key industries, the automotive industry being among the most obvious. For instance, the country's speed limits have not as yet been touched by any climate policy regulation, albeit the subject has been discussed at government level. Additionally, Germany is benefiting from an increased global interest in green technology products. Given that 'made in Germany' is regarded as a renowned international trademark, environmental and climate technologies already form a vital part of the economy. In order to compete in the regulatory arena with China and the United States successfully, Germany has a great interest in shaping global climate standards and in this regard considers the European Commission as a powerful ally.

By relying almost entirely on the expansion of renewables to achieve its 'Energiewende' and by phasing out nuclear power before coal, Germany has chosen a 'Sonderweg' (special path) compared to many European countries. It has relied on natural gas as a bridging technology. However, it is precisely this 'Sonderweg' that now poses challenges for Germany in view of the Russian war of aggression on Ukraine, given that before the war, 55% of Germany's gas was imported from Russia. The government agreed to stop the Nord Stream II pipeline project, which had previously been strongly contested in Europe, but only after Russia's invasion of Ukraine. The need to replace gas in Germany's energy mix has put former 'taboos' such as extending the lifespans of nuclear power plants or reactivating the coal sector back on the political agenda. Additionally, the Federal Minister for Economic Affairs and Climate Action, Robert Habeck, went to Qatar in order to negotiate liquified natural gas options, albeit building Germany's first terminal in Wilhelmshaven is only just underway. According to polls from June 2022, the German population supports such alternative considerations. Due to the Russian war against Ukraine, the issue of energy supply security has become the federal government's top priority.

The German 'Sonderweg' also impacts European climate policy, with the EU taxonomy being the most recent example, reflecting both France's preference for nuclear energy and Germany's focus on natural gas. Accordingly, Germany has the potential to put its own stamp on EU climate policy. It is also a strong actor in promoting more European integration in the field of energy policy. Given the looming gas supply crisis, the German government has called for solidarity among EU Member States and recently signed a respective Memorandum of Understanding with Austria, Czechia, Hungary, Poland and Slovakia.

Generally speaking, Germany can play a pioneering role by taking decisions which generate signals at European level. The Federal Constitutional Court called on the government to advocate even stronger climate protection at international level. Furthermore, the new government comprising Social Democrats, the Greens and the Liberals—the so-called traffic light coalition because of the parties' political colours—communicated its willingness to act as a motor for further European climate initiatives. Climate protection is at the heart of a coalition agreement that defines significantly more ambitious targets than those of the previous coalition. The traffic light coalition has also introduced a 'climate check' for all draft legislation. Traditional climate ministries such as the Federal Ministry of Food and Agriculture and the Federal Ministry for the Environment, Nature Conservation, Nuclear Safety and Consumer Protection, are in the hands of the Greens. Additionally, the Federal Ministry for Economic Affairs and Climate Action and the Federal Foreign Office, that are also headed by ministers from the Greens, have been given a strong focus on environmental and climate policies. Consequently, compared to the early years of climate protection in Germany, the fight against climate change has become a major cross-cutting issue in German politics.

Recommendations

This momentum should be used to spread still further the narrative that climate change can be fought only collectively in order to expand European integration. As a long-term aim, the EU could strive to create a European electricity market, for which Germany has already offered itself as a driver. Furthermore, the EU should closely link debates on Europe's strategic sovereignty to the field of energy and create a market for materials needed in the construction of wind and solar power plants, given that currently production is heavily dependent on China. The German government should make use of momentum resulting from the latest election results to combat climate change in a holistic manner rather than almost exclusively focusing on the 'Energiewende'. It should not only develop ambitious and realistic climate protection programmes, but also urge the European Commission to strengthen existing initiatives where necessary. Last but not least, it is precisely with a Green Foreign Minister that Germany has a real opportunity to live up to its reputation as a pioneer of climate protection at international level and advance climate diplomacy as well as rapid implementation of the Paris Agreement.

Funda Tekin is director at Institut für Europäische Politik (IEP), External Senior Research Fellow at the Centre of Turkey and European Union Studies (CETEUS), University of Cologne and TEPSA board member.

Since its founding in 1959, IEP has been a non-profit organisation dedicated to the study of European integration. It is one of the leading foreign and European policy research centres in Germany, serving as a forum for exchange between academia, politics, administration and political education. IEP's mission is to apply scholarly research to issues of European politics and integration, propose ways forward and promote the practical implementation of its research findings. IEP is also a member of TEPSA.

Christina Goßner was a project assistant at Institut für Europäische Politik (IEP).

Since its founding in 1959, IEP has been a non-profit organisation dedicated to the study of European integration. It is one of the leading foreign and European policy research centres in Germany, serving as a forum for exchange between academia, politics, administration and political education. IEP's mission is to apply scholarly research to issues of European politics and integration, propose ways forward and promote the practical implementation of its research findings. IEP is also a member of TEPSA.

Greece's New Energy and Climate Strategy. A Story of Hope?

Angeliki Papantoniou

Setting the Scene

Climate Change is a serious concern for Greek citizens. According to a survey conducted by Focus Bari and YouGov, a staggering 95% of Greeks are in no doubt that the climate is changing due to anthropogenic factors, believing that this change poses a grave threat to humanity.

Greece's total energy consumption per capita is 33% below the EU average, with renewables accounting for around 22% of its total energy use. Looking at the survey in more detail, 52% of the participants see 'a lot' of changes in the climate, while 33% see 'quite a lot' and only 5% deny the phenomenon. More than 60% of Greeks blame climate change on human activity, with industrial and developing countries, such as China and India, receiving the greatest blame, while all respondents hold the governments, both local and central, accountable for dealing with the impact of climate change. Moreover, 71% of respondents believe that climate change and extreme weather phenomena have a negative impact on physical and mental health, while 77% believe that all consumers can make a positive impact in regard to reversing any negative consequences.

Within this context, Greece's new energy and climate outlook has more recently been shaped by the current upheaval created by Russia's invasion of Ukraine, which further exacerbates already challenging economic and social consequences from the COVID-19 pandemic and the underlying economic crisis going back to 2008. While Greece does not share the devasting fate of Ukraine, it still faces certain serious consequences, especially in relation to its energy security, with energy (and food) prices soaring to unprecedented highs. While the Greek government was able to provide aid packages totalling EUR 4 billion to vulnerable groups in 2021, a long-term strategy to lead Greece out of the global energy crisis is now vital.

A. Papantoniou (✉)
Queen Mary University of London, London, Greece

Greece's Climate Change Performance

The Climate Change Performance index compares 60 countries with the EU in the areas of greenhouse gas emissions, renewable energy sources, energy use and climate policy. It provides a comprehensive overview of efforts in the countries it examines, evaluating to what extent they are on track to meet the Paris Agreement's global goals. Since last year, Greece has jumped ten places, an impressive ranking considering not only the combined challenge of the economic crisis and COVID-19, but also the fact that the EU has slipped six ranking places since the previous year. Signs of this improvement are demonstrated in the new Climate Change and Energy Transition legislation.

The New Climate Change and Energy Transition Legislation

On 27 May 2022, Greece passed its first National Climate Law with specific targets to fight climate change and eliminate coal in its energy generation—The National Climate Law: Transition to climate neutrality and adaptation to climate change (Law 4936/2022).

In terms of mitigation, the legislation sets interim targets for Greece to achieve at least a 55% cut in greenhouse gas emissions by 2030 and 80% by 2040, towards net-zero emissions by 2050. Moreover, the government is planning investments of approximately EUR 10 billion to expand the Greek power grid by 2030 and has committed to speeding up development of renewables to more than double in the country's electricity production. In addition, this law introduces the establishment of a process for developing sectoral carbon reduction relating to seven sectors in the economy, a progressive tool that exists only in the most advanced European climate laws. More specifically, targets include a commitment to cut dependence on fossil fuels in energy production from 2028 onwards, emphasising particularly two of the environmentally worst pollutants, lignite and brown coal.

Mitigation aside, the law also establishes adaptation measures, by providing local plans and programmes for coping with climate change, under a broader National Adaptation Strategic Plan. Moreover, it provides for: progress indicators within the relevant targets; progress assessment and target adjustment procedures; together with the creation of a governance and participation system for climate action, in line with established international standards on environmental human rights.

This law is a step forward for the synchronisation of Greece with the greater international objectives of sustainable development and energy security. It protects the rights of future generations and children from the effects of climate change, thus indirectly committing to the Sustainable Development Goals.

However, there is still room for improvement. Greenpeace and other environmental groups have stated that whilst the legislation is an important albeit small step towards climate neutrality, nevertheless it did not provide for the necessary strategic steps that will enable Greece to move quickly from a fossil fuel-based energy supply.

Moreover, Green-Tank, an environmental think-tank, which participated in the law's legislative process at various stages, noted that compared to the bill submitted for public consultation in November 2021, the level of ambition had been lowered in a number of areas. This contradicts the global and EU regional energy and climate efforts, laid down in the Fit for 55 package from July 2021, which included the European Green Deal, targeting neutrality by 2050, with an intermediate step of 55% by 2030. Bearing in mind specifically energy security because of the war in Ukraine, the REPower EU plan by the European Commission, which provides for energy saving and clean energy production, aims to make Europe 'Independent from Russian fossil Fuels' well before 2030.

Enhancing Energy Cooperation in Response to the War in Ukraine

Greece has been instrumental in being an energy sector stabilising agent since the beginning of the war in Ukraine by intensifying its regional efforts in the gas sector. While gas raises some environmental concerns, it remains a viable short-term option for energy transition.

Firstly, the new pipeline 'Interconnector Greece-Bulgaria' will supply gas from Azerbaijan, which Greece receives via the Trans-Anatolian pipeline that goes through Caucasus and Turkey. Bulgaria has already agreed to be part of the system, thereby abandoning its Russian gas supply.

Within the same context, Greece, Cyprus and Israel have embarked on an ambitious cooperative project, the Euro-Asia Interconnector, which is expected to be completed by 2024. The European Union aims to cut its reliance on Russian gas by two-thirds this year and end all Russian fossil fuel imports by 2027 at the latest. Cyprus, Greece and Israel have all agreed to build the world's longest and deepest underwater power cable, which will traverse the Mediterranean seabed and connect with electricity systems.

This enhancement of cooperation, while now solely focusing on gas, could potentially expand Greece's role as a key player in energy security, more broadly to include renewable energy and other climate change initiatives. At the heart of climate change, mitigation and adaptation measures are cooperation between states and this development can certainly lead to regional climate action sustainable development progress.

Over the coming months, Greece and the EU need to face a number of key issues. These include the greater impact of energy security and its connected economic instability, a bolder renewable energy strategy and the further integration of climate change adaptation measures in what is now a heavily mitigation-focused climate strategy.

Angeliki Papantoniou is an environmental and human rights lawyer, with experience both in academia and practice. Her academic research focuses on environmental law and children's rights. Her monograph 'Children and the Environment: Concepts of Legal Protection' is due to be published this year. Her research interests also include climate change law, the right to health, indigenous people's rights, the concept of vulnerability in law and the law of the sea. She is the LLM. Course convenor for Climate Change Law and the International Law of the Sea at Queen Mary University London and has been a practicing lawyer in Greece with membership of the Athens Bar Association, in the areas of refugee law, environmental law and human rights law.

Queen Mary University of London is a leading research-intensive university and a unique place of world-leading research as well as unparalleled diversity and inclusivity, that lives and breathes its history and heritage and is embedded in the communities it serves.

Hard Pressed by External Actors: Sustainability Transition in Hungary

Dániel Muth and John Szabo

Climate change has recently become a salient social issue in Hungary, as numerous surveys and polls confirm that Hungarians are as concerned about the devastating effects of deteriorating environmental conditions as their European peers. The ruling Fidesz party, continuously in power since 2010, has also shifted rhetorically from a moderately climate-sceptic stance to a pro-environmental position, but it did so in a context where the country's carbon-intensity per capita is relatively low. Despite the populaces' similar levels of concern over climate change, there are immense differences between the action that EU15 countries (which include the 15 countries in the EU from 1 January 1995 to 1 May 2004) and Hungary plan to take. Hungarians may express strong ecological concerns, but their material considerations prevail, with the government frequently claiming that the country is already on a low carbon trajectory.

Climate policy is largely set by the government's political and economic cost-benefit analysis. It is reluctant to channel investments that would deviate from pre-existing energy policy goals, even if those would yield emission reductions, and frequently notes that Hungary should wait for technologies to mature, making the energy transition less costly. The government tends to underscore that the latter is paramount because of the country's low income in comparison to its Western European counterparts. Hence, it simply postpones action to evade politically damaging energy price increases and avoid weakening its pro-Russia foreign policy orientation. Such a dominantly cost-centred discourse and reactionary approach builds on the government's bid to position itself as the provider of low utility costs, which disincentivises investment in the residential sector, for instance in regard to energy efficiency or household-sized solar photovoltaics.

D. Muth (✉) · J. Szabo
Institute of World Economics (IWE), Centre for Economics and Regional Studies, Budapest, Hungary
e-mail: muth.daniel@krtk.hu

Hungary's path towards climate neutrality is already well paved. The electricity sector is relatively low carbon, because the Paks nuclear power station generates around half of the country's overall domestic output. The government has signalled a firm commitment to substitute this ageing plant with Paks 2—a project developed by Russian Rosatom which faces increasing headwinds as the war in Ukraine continues. A domestic solar boom—on the back of a net-metering system for households and a green premium for larger installations—is decarbonising electricity generation, with production levels rising virtually from nil to just over three gigawatts in 6–7 years. Whilst progress is set to continue, concerns linked to administrative barriers and grid capacities are mounting. Moreover, the outright ban on wind turbines skews the country's overall production portfolio, introducing challenges in balancing the grid, amongst other issues. 2030 targets are thus based on maintaining nuclear power's role, the expansion of solar photovoltaics and Hungary's undertaking a fuel source conversion at the relatively old Mátra Power Plant (the country's second largest next to Paks) from lignite to natural gas and biomass.

Hungary used 10.44 billion cubic meters of natural gas to meet a third of its primary energy demand in 2020, according to the Hungarian Energy Authority, just under half of which was consumed by households. Decarbonising this segment of the heating sector poses a real challenge for a number of reasons: natural gas consumption is widespread; the building stock is in poor shape; subsidised consumer prices have impeded any meaningful action; while geopolitical alignment between the ruling government and Russia prevents a quick shift away from the fuel. External funding will play an especially prominent role in driving Hungary's pivot away from natural gas in the heating sector and it may become just as important as in transportation. In the latter case, relatively low income per capita means that many consumers simply cannot afford electric vehicles and hence they are reliant on an ageing diesel fleet. This problem is exacerbated by Western European consumers 'dumping' old internal combustion-engined vehicles in Central and Eastern Europe.

Climate change represents a collective action problem requiring concerted efforts from all countries and leadership from EU institutions. However, the Hungarian government resists developing greater supranationalism through multilateral climate change mitigation efforts. This has driven a clash with EU institutions and the continual scapegoating of Brussels for increasing energy prices. In practice, the government opposed numerous ambitious measures initiated by the Commission and other EU Member States, including countering proposals for climate neutrality in the Union or launching a public campaign against the proposed EU Emission Trading System covering the transportation and residential sectors (under the premise 'No to climate tax'). High energy prices and the war in Ukraine have further underpinned the government's reluctance to take investment-intensive action in support of the transition, because it is interpreted and communicated as an unacceptably costly endeavour.

Reluctance to engage in serious decarbonisation efforts persists in Hungary because of a general aversion to shift away from the status quo. There is little political will from the government to substitute a general low energy price policy with more targeted measures that also support emission reductions by driving investments in multiple renewable technologies at scale. Moreover, there is little ambition to disentangle foreign policy from hydrocarbon and nuclear relations. The government frequently ignores recommendations from civil society and various green industry associations who suggest that taking action would not only enable decarbonisation, but also underscore critical issues, such as benefits easily outweighing costs, even in the short- to medium-term by reducing Hungary's trade imbalance. This would also bring about the creation of new jobs and improvements in air quality.

Recommendations

In response to the above, we articulate four key recommendations. Firstly, directly allocate EU funds to a number of local actors to support the execution of large-scale energy efficiency and fuel-switching programmes in the residential heating sector. These investments are essential for meeting climate targets. Any EU funds withheld from the national government due to a number of concerns, including corruption, simply delays the energy transition.

Secondly, revenue generated via the new Emission Trading System should support the transition and dampen pressure on the most vulnerable consumers. Energy poverty is already pervasive in the country and this needs to be tackled, which can be done through sweeping energy efficiency and complex housing programmes.

Thirdly, EU policy to decarbonise the transportation sector needs to consider the flow of emitting/polluting diesel vehicles into Central and Eastern Europe. The purchasing power of locals does not support a shift to electric vehicles without subsidies, which need to be provided in a just manner, added to which alternative modes of transportation need to be heavily subsidised (for instance, bicycle or rail infrastructure).

Fourthly and finally, the Russian war in Ukraine has shown how prominent the cost of energy is in Hungarian politics. Political support and lucrative business ventures hinge on a Russia hydrocarbon-based energy system. This underscores and amplifies the barriers within local energy transition. However, by shifting the discourse and underlining the nexus between climate, energy security and the increasing affordability of renewables, public support can be harnessed to exert further political pressure.

Dániel Muth is a PhD Candidate at the Doctoral School of Political Science, Public Policy and International Relations, Central European University, Vienna and a Junior Fellow at the Institute of World Economics (IWE), Centre for Economic and Regional Studies, Budapest. His research focuses on the political economy of global carbon pricing mechanisms.

IWE is an independent fully-fledged research institute in Hungary dedicated to analyse trends and mechanisms shaping the global economy. A priority area of ongoing research efforts focuses on the embeddedness of energy transition and climate policy developments in the surrounding social systems. IWE is also a member of TEPSA.

John Szabo is a Research Fellow at the Institute of World Economics (IWE), Centre for Economic and Regional Studies, as well as an Assistant Lecturer at the Department of International Relations and European Studies, Faculty of Social Sciences, Eötvös Loránd University. His interests primarily lie at the intersection of energy and social relations.

IWE is an independent fully-fledged research institute in Hungary dedicated to analyse trends and mechanisms shaping the global economy. A priority area of ongoing research efforts focuses on the embeddedness of energy transition and climate policy developments in the surrounding social systems. IWE is also a member of TEPSA.

'Now, We Need Action': Ireland's Fight Against Climate Change

Luke O'Callaghan-White

In May 2019, Ireland declared a climate and biodiversity emergency, becoming only the second country—after the United Kingdom—to do so. This parliamentary declaration followed the publication of a report on climate action from a joint-parliamentary Committee. The Committee produced a broad suite of measures and recommendations for Government to apply, in order to establish a 'new climate policy architecture' and a 'comprehensive governance framework' for the necessary transformational changes to quell the climate crisis. Hildegarde Naughton, then Chair of the Climate Action Committee, welcomed the declaration as an important statement, but emphasised that 'now, we need action'.

The 2020s: A Decade of Ambitious Emissions Reduction

Following the general election in 2020, a three-party coalition was formed in July of that year. There was a strong green dimension to the government programme, which was released during the COVID-19 pandemic's first wave. This new coalition committed not only to a 51% reduction in overall greenhouse gas emissions by 2030, compared with 2018 levels, but also to carbon neutrality no later than 2050. This pledge was then enshrined into Irish law. Ireland's emissions reduction target for 2030 is one of the most ambitious commitments in the world and was generally understood to convey the government's aim to address this emergency and transform Ireland from a climate 'laggard' into a climate 'leader'.

However, a report published in June 2022 by the Environmental Protection Agency shows that there is a significant implementation gap between current climate policies and those needed to meet the 2030 commitments. There is a growing

L. O'Callaghan-White (✉)
Institute of International and European Affairs (IIEA), Dublin, Ireland
e-mail: luke.ocallaghan-white@iiea.com

concern that the government will miss its 2030 target, just as it has missed every previous emissions reduction objective.

Public Support for Climate Action in Ireland

There is certainly broad public support for robust climate action measures in Ireland. A Eurobarometer survey, conducted between March and April 2021, found that 94% of respondents in Ireland think that the EU economy should be climate-neutral by 2050. This is 4% higher than the EU average.

In October 2021, Friends of the Earth published an opinion poll—conducted by Ireland Thinks—which showed that a significant majority of the Irish public supports the Government's 2030 emissions reduction target. In the media framing of climate action in Ireland, significant attention is placed on the so-called 'rural-urban' divide, where rural Irish people are less supportive of tackling the climate emergency than their urban counterparts. While the polling indicates that support for climate action is indeed strongest in Dublin and among young people and women generally, the regional divide is less than might be expected. For instance, 79% of respondents in Munster and 72% in Connacht-Ulster—all predominantly rural provinces—support the Government's 2030 target. This compares with the national figure of 80%.

The Winds of Change

Ireland's path to carbon neutrality will be powered by wind. To meet the expected increase in electricity demand and ensure that the proportion of renewable electricity reaches 80% by 2030, the Government will need to harness the untapped potential of wind energy off the island's west coast. At present, Ireland is the only EU country with an Atlantic coastline which has yet to develop an offshore wind industry.

Irish policy-makers highlight that exploiting Ireland's very favourable offshore wind regime will not just be essential for delivering Ireland's decarbonisation targets but will also allow for the electrification of other sectors, such as transport and heat, thereby concomitantly reducing emissions in these sectors.

Wind energy also presents significant social and economic opportunities for people in Ireland. The Sustainable Energy Authority of Ireland estimates that further expansion of offshore and onshore wind could create 20,000 jobs in the installation of infrastructure, operations and maintenance by 2040.

The need to accelerate transition to renewable energy sources in the EU has been firmly underscored following Russia's invasion of Ukraine on 24 February 2022. However, Ireland is not reliant on Russian gas to meet its energy needs. Approximately 25% of Irish gas is supplied from indigenous sources, while the remaining 75% comes from the UK, which is itself also not reliant on imports of oil or gas from Russia to meet its energy demands.

However, Ireland is not immune to the upward pressures on wholesale gas prices and the knock-on effects this has in triggering electricity price hikes, concerns which have helped to illuminate the economic potential of Ireland's clean energy sources. Ireland's offshore economic area is several times larger than the island itself; accordingly, through an ambitious and effective offshore wind strategy, it can become a net exporter of clean electricity, ultimately contributing in European terms to the twin aims of clean energy generation and accelerated divestment of Russian energy imports.

Managing Electricity Demand in the 2020s

There are significant obstacles that must be addressed and overcome if Ireland is to live up to its commitment of achieving 51% emissions reduction by 2030 and climate neutrality by 2050. One such issue is the management of electricity demand. As Ireland's economy decarbonises, an increasing number of economic sectors will become electrified; inevitably, electricity demand will also grow.

While ramped-up offshore wind generation will increase capacity to meet the increased levels of electricity demand, there is nevertheless the potential for severe power system strain. This threat is enhanced by the presence of major multinational companies such as Amazon, Meta and Google, who base their storage facilities in Ireland. While there are fewer than 100 data centres operating in the jurisdiction, data centres consume a significant amount of electricity. Reports from the Central Statistics Office show that in 2021, data centres used a greater proportion of electricity than the total usage for all rural households in Ireland. Moreover, the overall consumption of electricity by these data centres has tripled over the last 6 years, to the extent that EirGrid—Ireland's national grid operator—forecasts that by 2030 their combined usage will make up 23–30% of total electricity consumed.

This extremely energy-intensive nature of data centres poses a risk to the proper functioning of Ireland's electricity network at a time when an increase in overall demand for electricity is to be expected, following the state's decarbonisation agenda.

Addressing Agriculture

The agri-food sector is one of Ireland's largest industries. In 2020, it accounted for almost 7% of modified gross national income and approximately 164,400 jobs, representing 7.1% of total employment. Significantly, the agriculture sector also contributes most to greenhouse gas emissions, in 2020 being responsible for over 38% of the country's total emissions, mostly from methane. For comparison, the European average for emissions from agriculture is 11%.

In line with the Government's Climate Action Plan—the programme of policies to achieve progress towards the 2030 emissions reduction targets—methane emissions will need to decline by almost 30%, to achieve an overall reduction of

22% in agriculture-related emissions by 2030. How this target is to be reached remains unclear.

Reducing emissions in the agriculture sector has proven difficult, with only a 2.3% reduction across the EU having been achieved since 2005. Given the degree to which agriculture contributes to Ireland's greenhouse gas emissions, Irish policy-makers and sectoral leaders will need to set out a clear pathway to reach the reductions target. At present, change is moving in the wrong direction. Under current policy measures, the country's Environmental Protection Agency still project that emissions in this sector will increase by 1.9% during the current decade.

Climate Change and the Irish-EU Relationship

The implications of runaway climate change will be unambiguously catastrophic for the planet. However, as Ireland and the EU reappraise economic orthodoxies to become carbon neutral by 2050, there is an opportunity to recalibrate and strengthen policy coordination.

In 2022, Ireland remains an energy island, poorly connected with other electricity grids and not yet connected with an EU grid, albeit the Celtic Interconnector linking France and Ireland will be completed later this decade. Through its Connecting Europe Facility, the EU should, therefore, consider investing in further grid interconnection, this time between Ireland and the rest of the EU. New interconnection infrastructure would not only help to integrate the Irish electricity grid with the rest of the Single Market, but would also provide a dependable source of renewable electricity for the EU.

Recommendations

During transition to a net zero future, it is imperative that neither the Irish Government nor the EU loses sight of justice's importance in climate and energy frameworks. Beyond important ethical considerations, if policy-makers introduce measures to reduce emissions which lack a sense of fairness in terms of public perception, this may foster social resentment and, in turn, serve to undermine the credibility of the Government or EU as a leader on climate action, ultimately slowing transition to a net zero future.

European policy-makers should prioritise the development of a European Supergrid. A coordinated, long-term approach to continental grid planning is necessary to integrate the level of renewables that will be required by 2050. This Supergrid would connect the best renewable sources to demand centres across Europe in an efficient and carbon-neutral manner.

Irish policy-makers should develop, as a matter of urgency, a national hydrogen strategy. Given its capacity to store energy, hydrogen can play an important pvotal role in addressing the issue of intermittency with renewables, as such making Irish wind energy generation a cheap, clean and dependable energy source. The storage potential of hydrogen is particularly beneficial for power grids, as hydrogen allows for renewable energy to be kept not only in large quantities, but also for long periods. This means that hydrogen can help improve the Irish energy system's flexibility by balancing out supply and demand. A roadmap for how best to optimise the green hydrogen economy in Ireland will be essential in fulfilling its long-term growth strategy which will, at least in part, be based around a capacity to export green hydrogen to continental Europe.

Luke O' Callaghan-White is the senior researcher at the IIEA. He leads the Institute's research on climate and energy policy. Luke's research interests include the geopolitics of clean energy transition, the nexus between climate change and security threats and climate diplomacy. Prior to joining the IIEA, Luke worked at the Toronto-based research institute, Canadian Centre for the Responsibility to Protect (CCR2P). Luke holds an MA in Political Science from the University of Toronto and a BA in Political Science and Geography from Trinity College Dublin. Luke is also a committee member of the Irish Centre for European Law, where he contributes as an expert in European energy developments.

The Institute of International and European Affairs is Ireland's leading international affairs think tank. It is an independent, not-for-profit organisation with charitable status. Its aim is to provide a forum for all those interested in the EU and International affairs, not only to engage in debate and discussion, but also evaluate and share policy options. It is also a member of TEPSA.

Vulnerable and Unprepared: Assessing Italy's Path to Fight Climate Change

Margherita Bianchi

The core of this chapter has been written BEFORE the change of government in Italy. Italy is located in an area which is particularly vulnerable to climate change. Indeed, in this regard the Mediterranean is considered to be a hotspot and as a result Italy, together with other Southern European countries, will be very heavily impacted. Here climate change is associated with rising temperatures, changes in rainfall patterns and prolonged periods of drought. Moreover, an increase in the frequency of extreme weather events is 9% more likely than it was 20 years ago, according to the Euro-Mediterranean Centre on Climate Change. All sectors of the economy are negatively impacted, with a particularly worrying trend being seen in agriculture. Changes are presenting risks for the availability of water resources and hence revealing the inadequacy of current infrastructures and adaptation strategies. The costs associated with climate change impacts increase exponentially as temperatures rise in different scenarios, with some forecasts indicating a loss of up to 8.5% by 2050 of gross domestic product. Climate change also interacts with socioeconomic and land use changes, as well as widening the economic gap between regions. Mitigating and adapting to climate change is thus an existential priority for Italy. The country is already suffering and will increasingly suffer high climate costs, for example in the agricultural sector, but despite this there is still a shortage of climate insurance.

Following a recent avalanche of ice in the Italian Alps—and the human tragedy that followed—as well as a number of recent extreme events (very high temperatures, violent hailstorms), climate change has made a return to general conversation amongst the Italian public, albeit as yet not assuming any priority on the political agenda. Draughts in the Marche region in the summer caused irreversible impacts to the population and territory. Over the spring and summer of 2022,

M. Bianchi (✉)
Istituto Affari Internazionali (IAI), Rome, Italy
e-mail: m.bianchi@iai.it

land temperatures have been reaching record levels, which has prompted debate about the country's limited hydroelectric production. This is also worrying from a security perspective, given the dramatic energy crisis that Italy is undergoing. However, despite all these signs, climate risks are still being underplayed in political debate nationally. While we discuss and drive a net zero process, we maintain a constantly conservative view in regard to climate risk estimates. Associating 'green' with 'costs' is likely to provide openings for populist arguments, thereby limiting the possibilities of success from a political and policy perspective. Because of green parties' weak role in Italy compared to other major EU countries such as Germany and France, any debate on climate change is mainly led by centre-left parties, although their narrative and approach are far from incisive on this topic. However, many civil society organisations, non-governmental organisations and sectoral associations, for instance in agriculture, are now more regularly raising climate change issues in the debate. In the run-up to political elections—set for 25 September 2022—climate change was still underplayed in parties' electoral campaigns. The winning party, Fratelli d'Italia, and other conservative parties forming the new government majority, traditionally consider the environment a peripheral issue.

Italy's energy mix continues to rely very strongly on fossil fuels, especially gas, which is also in line with past political choices, for instance the 2011 referendum on nuclear energy. The energy sector is unquestionably the largest contributor to total national greenhouse gas emissions with a share of 80.5% in 2019. Other major emission sectors are industrial processes and product use (8.1%), agriculture (7.1%) and waste (4.3%). In recent years, though, the country has ambitiously embarked on its journey to cut its greenhouse gas emissions. Indeed, Italy has achieved overall reduction at a faster pace than the EU average and across all economic sectors over the 2005–2019 period. The country has seen a number of legislative and strategic policy updates related to climate action and now, in line with EU partners, it backs the European Green Deal vision and a green recovery out of the pandemic. This recovery and resilience facility, with its green strings attached, is particularly valuable for Italy (as one of main beneficiaries) in enhancing its ability to make a significant contribution to implementing the European Green Deal.

The Italian recovery plan is part of a wider framework of incentives and reforms promoted by the former Ministry for Ecological Transition to reach the 2030 and 2050 objectives, among which are mechanisms to support renewables or initiatives for the conservation of parks and biodiversity. Italy expects not only to triple its production of solar energy and double its production of energy from wind by 2030, but also to phase out coal by 2025. Indeed, the Italian National Energy and Climate Plan identifies significant cost reductions from wind and photovoltaics as key and will seek optimisation by upgrading existing installations. However, this plan now needs to be updated in light of recent amendments contained within the European Climate Law. This concerns the Recovery plan, particularly the many pledges which Italy agreed at the UN Climate Change Conference (COP26) in November 2021 and COP27 in November 2022, let alone the current circumstances in Ukraine and decisions that will be taken to respond to this crisis. Various temporary measures, such as an increased use of coal-fired power plants for electricity generation should there be a sudden halt of Russian gas supplies, are being considered as potential emergency solutions. In the meantime, Italy is trying to diversify its supplies, an issue which involves difficult choices that can affect our energy infrastructure in the medium term.

From 2021 soaring natural gas prices have pushed the Italian government (alongside many others) to subsidise energy bills. Even before Putin's invasion of Ukraine, reconciling environmental sustainability with energy security and competitiveness has been by no means a simple exercise in our country. Former prime Minister Draghi grasped the complexity of this challenge, establishing himself as a convincing interlocutor in Europe and relaunching cooperative multilateralism on the climate front despite growing international tensions. These developments were also facilitated by the leading role played by Italy throughout its presidency of the G20 and joint organisation of the 26th Conference of the Parties (Cop26) with the United Kingdom (2021). Moreover, decisive government action is very much in line with Italian citizens' growing concerns about the climate emergency, which is demonstrated by the public's increasing interest in any policies being undertaken by the government on this issue. Many citizens (66% according to a recent IPSOS survey) seem to agree on the necessity of corporate action to combat climate change and believe (71%) that individuals also have their own responsibility to fight it. However, according to some other surveys, climate change is still perceived by part of the population as a distant problem, particularly citizens in the South. Following the political crisis that led to Draghi's resignation, uncertainty over Italy's role, willingness and strength to fight against climate change both at home and abroad, remains high, especially after the rise of conservative Meloni as new PM. At COP27 Meloni merely followed up former decisions taken by the Draghi government but without any relevant leadership on the most hot topics on the agenda.

Italy needs reforms not only to make the most of available funds, but also to attract green investments, besides working to gain and maintain consensus on the side of citizens. Of the EUR 5.5 billion made available by the 2014–2020 European Regional Development Fund for energy, climate and environment projects, Italy has spent only 58%. Furthermore, Italy is struggling to present itself as a stable, competitive and reliable economy in the global arena. Hence, there is a crucial need for reforms that the country will have to carry out going forward, such as simplification and justice above all, to attract (green) investments. Moreover, outdated regulations, slow permit issuance, discretion in environmental impact assessment procedures, blockades by superintendencies, uneven regional regulations, disputes between institutions, all significantly slow down the renewables processes in Italy. Protection of landscape value, a strong 'not-in-my-backyard' (NIMBY) syndrome and risks of economic unsustainability also hinder the progress of Italian renewables initiatives. Hence, from Italy's perspective, acceleration in renewables permits in the context of the EU strategy 'REPowerEU' is to be welcomed.

To overcome the compartmentalisation of climate policy in order to create a bridge between rhetoric and practice, Italy has been trying to embed climate and energy discussions in politics and policies, but we are still only at the beginning of this process. Draghi's government pushed forward some distinctive institutional reshuffles to tackle the climate crisis – now again in some cases turning back with the new government The Ministry of Ecological Transition as well as the Ministry of Infrastructure and Mobility Sustainability established by Draghi have already been renamed by Giorgia Meloni with more traditional names, such as, in the first case, Ministry for the Environment and Energy security. A special envoy was established to represent Italy in international climate negotiations.

The outbreak of war in Ukraine has unexpectedly and worryingly heralded a turning point for the EU, within which Italy is facing a complex set of challenges regarding its energy governance and the need to maintain secure, sustainable and affordable supplies. Italy is one of the most exposed countries to potential disruptions due to reliance on the Ukrainian gas corridor and the relevant role of gas in its energy mix. Today, 50% of power generation is produced by gas power plants. In this context, Italy has been working to ensure adequate supply with new and immediate alternatives. The first response has been to use established alliances with producing countries to find replacement sources of energy for European needs, in particular through the Italian oil and gas energy company Eni. Building capacities and win-win alliances in neighbouring countries and beyond are though crucial (and in our interests), hopefully even with some green strings attached, to respond to these urgent security needs. The Italian government, for example has expressed its intention to work with Algeria on the production of green hydrogen and although this association in its infancy, such conversations are moving in the right direction. Giorgia Maloni's action is in continuity with that of Draghi for what concerns the tackling of this energy crisis. Less emphasis unfortunately have been put for the moment on the role of RES also in security terms, by both Draghi and Meloni.

In light particularly of the war in Ukraine, but also with a view to last year's energy crunch, President Draghi supports the progressive strengthening of Europe's integration process in order to improve the EU's ability to cope with the crises it is facing. Until now, many of Italian proposals about how the energy market should function—for instance, the gas price cap—have been taken into account and as such this request is being evaluated by the EU executive. At last, an agreement on the price cap was found in December 2022. The government has already intervened (and continues to do so with Meloni) to mitigate the effects of increasing energy prices on the most vulnerable households and there has also been action to help businesses. Less political capital was spent by both governments on other european proposals, such as the joint purchasing. The country needs to develop a more serious message with regard to savings campaigns and energy efficiency as Putin continues to weaponise energy, especially to establish a secure energy framework in 2023-2024-2025. This is absolutely crucial.

Recommendations

Public awareness of how serious the effects of climate change must be developed and increased at local levels. Indeed, a change in the narrative at both national and European levels, focusing on the cost of a 'business as usual' policy, is of paramount importance in order to build a consensus for delivery of the country's energy transition. Italy should undertake further much needed work on renewables' permits and the social dimension of new clean infrastructures. Climate risks must be better integrated into national political debate.

As climate change accelerates, adaptation cannot remain a secondary priority for Italy and the EU. Regarding the current crisis, the Union should find a fiscally, socially and environmentally acceptable way of helping citizens and industries face high energy prices as well as assisting Member States in taking the right decisions. Whilst there is clearly an urgent need to reduce and ultimately eliminate our use of fossil fuels, this makes no short-term strategic sense in times of energy crisis created by Russia's actions.

Renewables and energy efficiency must be accelerated by all means and in all sectors. Moreover, this needs to become the guiding pillar of all energy/climate policies as it supports every dimension of the energy trilemma—security, sustainability and affordability. At the same time, energy savings in terms of the current crisis are fundamental. Italy and the EU have to work on the demand-side through awareness campaigns and measures.

Stronger integration is needed within the internal energy market not only to facilitate responses to the current security problems we are experiencing, but also to make possible speedier decarbonisation. Any lack of solidarity and harmonisation of rules takes us in the opposite direction. Italy and the other Member States should understand this and moreover the EU should continue providing guidance for better integration.

Margherita Bianchi is Head of the Energy, Climate and Resources Programme at the Istituto Affari Internazionali (IAI). She is a member of the scientific committee of the online magazine AffarInternazionali. She deals in particular with the geopolitics of energy transition and analysis of the European Green Deal, having previously had experience with the European Parliament and the Task Force of the Italian G7 Presidency. She graduated with honours in Political Science and International Relations from the Catholic University of Milan, before specialising in European Affairs and EU Law with two master's degrees from the LUISS School of Government in Rome and the Institut d'études européennes (IEE-ULB) in Brussels.

IAI is a private, independent non-profit think tank, founded in 1965 on the initiative of Altiero Spinelli, to promote awareness of international politics and contribute to the advancement of European integration as well as multilateral cooperation. IAI is also a member of TEPSA.

importance, in order to build a consensus for delivery of the country's energy transition. Italy should undertake further much needed work on renewables' permits and the social dimension of post-fossil infrastructures. Climate does not be later mitigated if past mistakes prevail.

Nature Friendly Latvia Against Its Unnatural Climate Change Problem

Aleksandra Palkova and Karlis Bukovskis

The Latvian government did not include climate change on the national agenda until 2018. The country's energy balance, decades-long societal support for nature preservation, significant numbers of eco-friendly government policies and geographical location all resulted in a country not seeing climate change as an immediate problem. The increasing political and social popularity of climate change issues, both in the EU and globally, finally caught up with Latvia only after Prime Minister Krisjanis Karins demonstrated personal political leadership on the matter.

Once the European Commission had declared its target of climate neutrality by 2050, Latvia understood that its moment had come to be among the best. Due to its extensive use of renewables in the energy sector, high forestation levels and small industrial production volumes, the country could eventually become a champion in fighting climate change. Not only have the OECD and the United Nations Framework Convention on Climate Change highlighted Latvian progress, but other countries are also already learning best practices and technologies from Latvia, whose society historically is steeped in a culture of environmentalism. The country's political leadership defines climate change as a 'golden opportunity for the Latvian economy' and the chance to become a leading EU country.

There are two striking examples of Latvia's environmental credentials. For instance, during the period of Perestroika, environmentalist organisations were the first civil society groups to be formed. Indeed, it was the Environmental Protection Club that in 1988 laid the foundation for establishing a mass national independence movement, the Popular Front of Latvia. Additionally, since 2008 hundreds of thousands of Latvians have been taking part in 'The Big Cleanup'—annual civic activity related to cleansing forests and meadows from garbage—action that is growing in popularity beyond Latvian borders.

A. Palkova (✉) · K. Bukovskis
Latvian Institute of International Affairs (LIIA), Riga, Latvia
e-mail: aleksandra.palkova@liia.lv; karlis.bukovskis@liia.lv

M. Kaeding et al. (eds.), *Climate Change and the Future of Europe*, The Future of
Europe, https://doi.org/10.1007/978-3-031-23328-9_16

Today's Latvian politicians and decision-makers are active in climate change. Latvia became active in implementing the European Commission's Green Deal policy and unreservedly adopted the Climate law. Furthermore, it became one of the first countries to diversify the energy sector and also follows Fit For 55. A number of amendments were also made to Latvia's strategic and policy planning documents, marking changes in the public administration's attitude towards climate change. For instance, in 2019 climate change was mentioned as a 'threat' to Latvia's security in its National Security Concept. Furthermore, government and various ministries express regular concerns about the climate both in statements and policy strategies. Green non-governmental organisations are active in Latvia, to the extent that an annual Green Innovation Award has been set up, by way of paying tribute to the greenest companies involved in environmentally friendly production and distribution.

The country's determination to use this 'golden opportunity' at EU level is deeply rooted in its historically 'green' energy mix. Latvia has the third-highest share of renewable energy sources for final energy consumption in the EU, with its total energy mix comprising 40% renewables, 32% oil and 21% gas. While the main source of greenhouse gas emissions in Latvia stems from the energy sector (excluding transport), the total amount is still lower than the EU average. By contrast, other greenhouse gas emissions are higher than the EU average: transport 30%, agriculture 25% and waste management 5%. Nevertheless, amongst all EU Member States, Latvia is considered to have one of the fastest declining greenhouse gas emissions rates when compared to 1990 levels. This more than 50% reduction is primarily due to a Soviet industry collapse in the early 1990s. Thanks to these achievements, Latvia has increased its national ambition to reducing greenhouse gas emissions by 65% and ensuring a 50% share of energy produced from renewable sources by 2030.

However, these aspects are exactly where the country's greatest paradox lies. On the one hand, Latvia's government has ensured that progress is continuing to be made not only in business, innovation and the green economy, but also regarding good greenhouse gas emissions statistics. On the other hand, whilst central government wants to advocate even more active policies, society in general does not see the need for any further efforts in tackling climate change, given that 'green transition' demands unaffordable additional investment. Moreover, society generally fails to make a distinction between national-level environmental problems and global climate change. In other words, while nature and the environment are essential to Latvian society, which is living in a land where forests cover more than half of the country's territory, at the same time, people generally do not see climate change as an immediate or acute problem.

For research purposes, Latvian society should be divided into three groups: firstly, there are those who deny that climate change is an issue, considering Latvia as green enough; secondly, there are some who are aware of general climate change trends, but regard them as unrelated to Latvia; thirdly and finally, others are strongly climate change-aware and seek to act accordingly. The Eurobarometer data makes it possible to conclude that this third group is about 5% to 7%, slightly up from previous years. As an additional point, according to domestic polling, Latvians generally see climate change as primarily an environmental problem without any clear correlation to the economy, infrastructure or energy sector.

Latvian society has often viewed the EU as a key 'source of inspiration' for policies produced by its government and in this regard climate change is no exception. Indeed, the government's reaction is viewed very much through an EU prism. In other words, if this is a relevant issue for the EU, then so must it be for Latvia. Latvians traditionally hold positive and favourable views about the EU. Since Russia's attack on Ukraine and the European Commission's launch of the RePowerEU initiative on energy diversification, the government has been trying very actively to convince society that change has now become a necessity.

It is clear that Latvia sees the fight against climate change as an opportunity to boost its international image, attract capital and deal with unfinished homework domestically. At the same time, society is not yet ready to see climate change as having implications beyond environmental problems. Latvian society does not regard the fight against climate change as an economic necessity, as a business opportunity or a chance to improve living standards. Rather there is a deep-rooted fear about additional expenses which will impact all Latvian households. One of the most visible financial concerns is to do with the purchasing of new electric cars in a country which has the EU's second oldest fleet, with each vehicle on average being more than 13 years old.

Recommendations

Taking all this into account, the following recommendations to the Latvian government are put forward.

Firstly, Latvian society must be better informed about the difference between state-level environmental protection and global climate change. Government strategies and cynical economic reasoning should be tangible and thoroughly explained to the whole of society.

Secondly, the nationwide 'green' narrative should be strengthened and instrumentalised if not weaponised. In other words, Latvia's 'green' mentality should be activated to gain wider public support for fighting global climate change.

Thirdly, Latvia needs to state its priorities in climate innovation, without trying to cover the entire field of industries.

Fourthly, enhanced collaboration with other Baltic and Nordic countries is needed to exchange views and ensure that the above-mentioned recommendations are necessary.

Finally, the European Commission must openly recognise the income differences not only among Member States, but also among people within the EU. Demands for a rapid shift to new technology are pointless while it is still too expensive.

Aleksandra Palkova is a researcher at the Latvian Institute of International Affairs (LIIA). She holds an MA in International Relations and is currently pursuing her PhD studies at the Riga Stradins University. Since 2021 Palkova has been an Associate Researcher at the European Council on Foreign Relations and a Senior Researcher at Riga Stradins University, where she studies security and climate policy issues.

LIIA is the oldest Latvian think tank that specialises in foreign and security policy analysis. This independent research institute conducts research, develops publications and organises public lectures as well as conferences related to global affairs along with Latvia's international role and policies. The LIIA is also a member of TEPSA.

Karlis Bukovskis is Acting Director of the Latvian Institute of International Affairs (LIIA) and Assistant Professor at Riga Stradins University. He is the author of numerous articles and scientific editor of several books on the European Union and International Political Economy. Bukovskis was a visiting Fulbright Scholar at the Johns Hopkins University SAIS in 2021.

The LIIA is the oldest Latvian think tank that specialises in foreign and security policy analysis. This independent research institute conducts research, develops publications and organises public lectures as well as conferences related to global affairs along with Latvia's international role and policies. The LIIA is also a member of TEPSA.

Climate Change Policies in Lithuania: As Usual, Words Speak Louder Than Actions

Ramūnas Vilpišauskas

The Rise and Fall. . . and Rise Again of Environmentalism's Political Salience

The environmentalist movement constituted a visible part of Lithuania's move towards independence in the late 1980s, by criticising Soviet industrial practices as unsustainable and campaigning against expansion of the Ignalina nuclear power plant, which at the time operated two reactors of the same type as those at Chernobyl. However, after the re-establishment of independence in 1990, political and societal focus shifted to transition reforms together with their economic and social consequences. Accordingly, this resulted in the decline of industrial activities and a growing share of services with lower pollution levels as a side effect.

EU accession revived the political salience of environmental issues, primarily due to the need for legal approximation of norms aimed not only at reducing pollution, but also making available EU funds for the transport and environmental sectors. This was presented initially in the form of an instrument for pre-accession structural policies and later as EU structural funds. In keeping with other EU Member States, Lithuania signed up to the Paris Agreement committing to a 40% reduction in greenhouse gas emissions by 2050 when compared with its 1990 levels. From a 2005 base, Lithuania has also committed to a 9% reduction in emission levels by 2030 (excluding sectors which participate in the EU Emissions Trading System). Furthermore, policy-makers have expressed support for a 55% reduction in greenhouse gas emissions by 2030 based on the country's 1990 levels, whilst at the same time expressing their concern that this would be a challenging task for those sectors which pollute the most, in particular road transport, energy and agriculture.

R. Vilpišauskas (✉)
Vilnius University, Vilnius, Lithuania
e-mail: ramunas.vilpisauskas@tspmi.vu.lt

M. Kaeding et al. (eds.), *Climate Change and the Future of Europe*, The Future of Europe, https://doi.org/10.1007/978-3-031-23328-9_17

It should be noted that political efforts to reduce dependency on Russian energy supplies have induced increased attention on the development of renewable energy sources such as solar and wind power. The country's share of energy produced from renewables reached 25.5% in 2019, thereby exceeding its 2020 target of 23%. Imported oil and natural gas accounted for about two-thirds of total energy supply, with their relative shares having increased since the end of 2009 when, in line with an EU accession treaty commitment, Lithuania closed its Ignalina nuclear power plant. Meanwhile its share of renewable energy sources used in the transport sector was only 4.33% in 2018, which is well below the country's 2020 target of 10%.

EU funds, including those from the National Recovery and Resilience Plan 'New Generation Lithuania' approved in 2021, have also been used to provide support for: developing renewable energy generation capacities, renovating residential buildings to improve energy efficiency and expanding the network of charging stations for electronic vehicles. The plan, which prioritises green spending by allocating EUR 935 million or 42.02% of foreseen funds, acknowledged that the most challenging areas are inefficient use of resources and high energy intensity. It aims to accelerate the renovation of residential buildings, long criticised for being too slow, and address the problem of growing emissions from road transport, especially cars.

Ambitious Political Statements…

Currently, political elites agree on the need for ambitious climate change mitigation initiatives in Lithuania. The conservative-liberal coalition government formed after the parliamentary elections in late 2020 dedicates considerable attention to climate change issues in its programme, with a declared intention not only to position Lithuania at the forefront of EU efforts in advancing its green deal, but also to support the most ambitious EU goals in this respect. Furthermore, it stresses that the EU's Green Deal will serve Lithuania's national security interests, as well as its energy and resource independence. Concrete objectives for the country in this area include recycled materials to be around the EU Member State average share by 2024, a 30% reduction in greenhouse gas emissions by 2030 compared with 2005 and climate neutrality together with a completely circular economy to have been achieved by 2050.

President Gitanas Nausėda elected by popular vote in 2019, has also prioritised environmental issues, following a campaign in which he promised to establish a welfare state. This was initially a vague concept, but later detailed by focusing on descriptions such as a just, green, innovative and secure country. He has also initiated a public relations campaign called Green Lithuania and proposed amendments to the law regulating legislative activities to include using impact assessment for new initiatives with respect to their potential effects on the environment and climate change. Furthermore, the President has publicly voiced his support for the government's ambitious intentions, especially in increasing the share of renewable energy sources and transforming other sectors in line with a green economy.

... But Decisions and Their Implementation Are Slow

Although Lithuania has been advancing relatively quickly in some areas, such as the use of renewable energy sources, elsewhere progress has been slow. This is reflected not only in having registered a relatively poor level of reduction in greenhouse gas from transport and agriculture, but also in struggling to develop more efficient uses of energy. Sometimes delays have resulted simply from difficulties in overcoming challenges associated with the coordination and practical implementation of policies involving many actors, as in renovating large residential buildings.

In other cases, political incentives to avoid any blame for increasing the population's financial burden could explain why there been some hesitation to introduce policy measures such as taxing pollution by cars. Rejection of a law presented by the Minister of Environment Simonas Gentvilas in mid-January 2022 to introduce a tax on vehicle pollution is a case in point. Some opposition figures in parliament argued that it was not the right moment to be increasing the tax burden on a population faced with high inflation. Indeed, although environmental issues receive increasing attention among the public, especially the younger generation, opinion surveys such as Eurobarometer show that in recent years shorter term fears about inflation have dominated concerns amongst Lithuanians generally.

Recommendations

To advance their climate change ambitions, policy-makers in Lithuania should not forget that in some areas, the state is still creating conflicting incentives for businesses and other sections of society by providing tax exemptions, for instance in the use of local fossil fuels, or subsidising activities such as farming that contribute to environmental pollution. Those types of incentives should be assessed with a view to their elimination at both national and EU levels.

The current inflationary environment, especially high fossil fuels prices resulting from a global mismatch between supply and demand plus Russia's war against Ukraine, create strong incentives for the population and businesses to change consumption patterns and search for ways of saving energy. State institutions should facilitate this process by providing information on how to save energy and shift towards more sustainable practices without upsetting price setting. The negative effects of high prices should be compensated for by targeted measures aimed at the most vulnerable households.

In addition, state institutions should also match their statements with activities, for example, by using environmentally friendly means of transport and taking into account environmental aspects in public procurement. Most importantly, there is an urgent need to improve the use of impact assessments to accompany new legislative initiatives and review existing laws with respect to their economic, social and environmental effects. Moreover, it would be advisable to arrange ex-post impact assessment of the rules regulating renovation of residential buildings so as to accelerate the process.

Finally, on top of EU and national funding, it is crucial that the regulatory environment should be conducive to innovations which help reduce pollution.

Ramūnas Vilpišauskas is Jean Monnet Chair (2020–2023) Professor at the Institute of International Relations and Political Science (IIRPS) at Vilnius University. From 2009 to 2019, he was Director of the Institute. In 2004–2009, he worked as Chief Economic Policy Advisor to the President of Lithuania, Valdas Adamkus. His main research interests include the political economy of European integration, together with policy analysis of public sector reforms and international political economy.

The IIRPS at Vilnius University is one of the most prominent social sciences institutions in Eastern Europe and the Baltic region. The Institute is an academic institution specialising in social and political sciences. IIRPS is also a member of TEPSA.

No Decarbonisation Without Taxation in Luxembourg

Guido Lessing

Ahead of the general election campaign in the autumn of 2018, the main Luxembourg parties (Democratic Party, Christian Social People's Party, Socialist Workers Party, Alternative Democratic Reform Party and The Greens) publicly signed an agreement to restrict the number of promotional campaign giveaways and campaign posters for the sake of environmental protection. At first sight insignificant, this agreement nevertheless demonstrates a certain degree of ecological awareness across party lines and attests to a level of interest in environmental issues among voters.

In 2018, as during the campaign for the European Parliament elections in 2019, blatant climate change denial was non-existent. However, one party from the Eurosceptic right-wing populist spectrum has been trying to sharpen its climate-sceptic political profile ever since. Unsurprisingly, in December 2020, its four Luxembourg Parliament members were alone in voting against the first national climate law, designed to comply with the Paris Climate Agreement of 2015, the law being passed with a majority of 56 out of 60 votes. Just two Members of parliament from the political Left abstained because they considered the law insufficient. Apparently, in order to boost its popularity ahead of the next elections in 2023, the Eurosceptic party is increasingly criticising the government's climate policy and its commitment to reducing CO_2 emissions. Its leading figures even go so far as describing human-induced climate change as a highly controversial claim in scientific circles. This standpoint leaves the party somewhat isolated among major decision-makers and stakeholders, ranging from the national business federation to climate activists, but seems to be of some appeal to its traditional supporters.

In his most recent State of the Nation Address in October 2021, the Luxembourgish Prime Minister, Xavier Bettel (Democratic Party—liberal) put the

G. Lessing (✉)
Luxembourg Centre for Contemporary and Digital History, Esch-sur-Alzette, Luxembourg
e-mail: guido.lessing@ext.uni.lu

M. Kaeding et al. (eds.), *Climate Change and the Future of Europe*, The Future of Europe, https://doi.org/10.1007/978-3-031-23328-9_18

fight against climate change as top priority on the list of challenges to be tackled, ahead of digitalisation, housing and social policy, albeit insisting that they are all inevitably linked.

According to a recent national poll conducted in May 2022, respondents seem to share the Prime Minister's view, but rank their priorities somewhat differently. In the current context, it appears that they are more worried about access to affordable housing (75%), increasing energy prices (62%) and inflation (59%), than the consequences of climate change (57%).

So, what factors will determine Luxembourg's path to climate neutrality? The Grand Duchy is the second smallest country in the EU with a total area of 2586 km^2 and no direct access to the sea. At the same time, Luxembourg has the highest population growth of all EU states, which puts additional pressure on the energy balance. In addition, at 13.24 tonnes per capita, carbon dioxide emissions are more than twice as high as the EU-27 average (5.91 tonnes).

Luxembourg's neighbours—Germany to the east, Belgium to the west and France to the south—are taking different paths towards the decarbonisation goal. With the French nuclear power plant of Cattenom within sight, just 22 km away from Luxembourg city, residents of Luxembourg are almost unanimous in their rejecting the use of nuclear energy. At the same time, the country is highly dependent on energy imports. Even if the share of domestically produced energy is increasing, it still covers just 19% of the country's needs, hence imports of most gas and electricity continue. Luxembourg and Belgium integrated their national gas markets in 2015, meaning that most of Luxembourg's gas imports transit via Belgium, whereas its biggest supplier of electricity is Germany. Hence, it is evident that the country's path towards decarbonisation cannot be taken without its European partners.

In order to achieve the 2030 target of cutting CO2 emissions by 55%, the National Climate and Energy Plan, adopted in 2020, focuses on increasing energy efficiency in the construction industry and promoting e-mobility. Policies on renewables primarily target wind and photovoltaics. Luxembourg's limited land area forces it to cooperate with European partners, as provided for in the National Energy and Climate Plan. As one of its flagship projects, in June 2021, the Ministry of Energy announced Luxembourg's involvement in the world's first offshore energy hub off the Danish coast, designed as an artificial island in the North Sea. Cooperation with Denmark will allow Luxembourg to include this 'green electricity' in its future CO2 balance so as to reach the EU targets.

When looking at final energy consumers, the transport sector accounts for by far the largest share, representing 62.2% of total consumption. The main reason for this is so-called 'fuel tourism', with cross-border commuters and residents from neighbouring regions benefiting from comparatively low taxation on fuels by filling up their vehicles in Luxembourg.

The question of whether or not setting society on the route to climate neutrality could be a tangible new narrative for European integration and cooperation is difficult to gauge. In the November-December 2021 survey by the European Investment Bank, only 10% of Luxembourg's residents felt that the EU was helping their country to tackle climate change, albeit that was before the start of Russia's war against Ukraine. Europe's vulnerability in terms of energy supply has created a growing awareness of the close link between fossil fuel consumption and security, although Luxembourg is less exposed to further energy cuts by Russia than other EU Member States. Nevertheless, in general terms, it is evident that the war will trigger an accelerated transformation to a more sustainable low-carbon society by reducing dependence on fossil fuels.

Recommendations

If the Union succeeds in organising a response to the external shock that is causing rising energy prices, EU citizens would also perceive Brussels as a major player in energy and climate issues. In Luxembourg specifically, where domestic resources are very limited and there is a high dependence on imports, this narrative would be even more true.

Given Luxembourg's limited space, cross-border initiatives to produce green energy should find substantial funding from the EU. This would alleviate the financial burden for cooperating countries and strengthen the common energy market. Moreover, Luxembourg needs to be part of a European hydrogen network in order to decarbonise industry and heavy-duty transport.

At the same time, there is an urgent need to reduce fuel consumption, since commuters who buy their fuel every day in Luxembourg make a disproportionate contribution to the country's energy footprint in comparison with other countries. There is a need to improve transport infrastructure with bordering regions and offer cross-border job tickets at unrivalled prices in addition to the free public transport Luxembourg introduced in 2020.

Finally, policymakers should be honest enough to tell citizens that energy consumption comes at a price for the environment and the wallet. It seems clear that energy consumption should be taxed more in order to finance technologies that increase efficiency and thereby bring about lasting lifestyle changes.

If we fail to do this as society, restricting electoral campaign giveaways will amount to nothing more than symbolic politics.

Guido Lessing is a Research Assistant at the Luxembourg Centre for Contemporary and Digital History (University of Luxembourg). He has many years of experience teaching history and civics in secondary school and has also co-authored various history and civics textbooks. After working for the Centre d'Etudes et de Recherches Européennes Robert Schuman (CERE) in Luxembourg, he joined the C2DH in September 2017. His main fields of interest are European integration and the history of Luxembourg in the twentieth century.

C2DH is the University of Luxembourg's third interdisciplinary research centre, focusing on high-quality research, analysis and public dissemination in the field of contemporary Luxembourgish and European history. It promotes an interdisciplinary approach with a particular focus on new digital methods and tools for historical research and teaching.

Malta and Climate Change: Balancing Opportunities with Limitations

Mark Harwood

While existing as a major issue for all countries, it has long been known that climate change represents a particular threat to small island states such as Malta, due to environmental sensitivity (caused by exposure to weather extremes and depletion of natural resources), economic vulnerability (due to isolation) and demographics (including migration and rapid urbanisation). When the European Commission published its Green Paper on Adapting Climate Change (2007), it listed six regions which are particularly vulnerable to the consequences of climate change, three of which (the Mediterranean basin, coastal zones and densely populated floodplains) applied directly to Malta. In this way, Malta, the EU's smallest and most densely populated Member State, has long been aware of its predicament, having experienced: a rise in mean annual ambient temperatures of 1.5 °C over the last 50 years; changes in precipitation patterns; as well as increased sea acidification and warming. As we will discuss, Malta has never had the luxury of being able to ignore the consequences of climate change and this is reflected in progress made to meet the EU's commitments towards climate neutrality.

Rapid Progress Hemmed in by Island Status

Malta is a group of small islands in the central Mediterranean, midway between Italy and Libya. With a population of half a million people (0.1% of the EU-27 population), it is the seventh most densely populated territory on earth, with few natural resources. The economy is dependent on tourism, service-orientated industries and the importation of most resources, whether energy, food or people (Malta has the

M. Harwood (✉)
Institute of European Studies, University of Malta, Msida, Malta
e-mail: mark.harwood@um.edu.mt

fastest population growth rate in the EU, reflective of its rapid economic expansion over the last two decades).

Malta generates less than 0.1% of the EU-27's total greenhouse gas emissions and has reduced its own emissions over the last ten years at a rate faster than the EU average; with a CO2 equivalent rate of 5.3 tonnes per inhabitant (2019), Malta has the second lowest per-capita emissions in the EU, well below 8.4 tonnes (EU-27 average). Much of the progress achieved in reducing emissions has been due to shifts in the energy sector during 2015, particularly the replacement of heavy fuel oil with natural gas in local electricity generation as well as the inauguration of an electricity interconnector with Sicily. In 2019 (most recent figures) final energy consumption of fuel was 52% oil and petroleum (primarily for transport), 43% electricity generated by natural gas and 5% renewables. Oil and petroleum imports largely derive from a mixture of countries, but natural gas for electricity generation is imported under an 18-year contract with Socar Trading, the trading arm of Azerbaijan's state oil producer.

According to the most recent data (2019), greenhouse gas emissions are the result of energy production (28%), buildings and tertiary industry (25%), transport (24%) (primarily air transport but also private vehicles, Malta being third only to Luxembourg and Italy in terms of the car to inhabitant ratio), industrial processes (12%) (primarily refrigeration and air-conditioning units), with manufacturing and agriculture representing 3% (or less). In trying to lower its emissions, Malta has focused on improvements in overall energy efficiency, through renewable energy, changes to the transport system and improved agricultural practices. Furthermore, in 2021 the government published its Low Carbon Development Strategy, with the focus on policies to extend free public transport, the electrification of private and public transport, promotion of remote working as well as 'nearly' zero-energy building standards.

The government has consistently reiterated its commitment to climate-neutral policies, with the Maltese Prime Minister stating in 2022 that whilst Malta might be a small polluter, the island still needed 'to do its part'. This was an important statement as government is the determining factor in combating climate change. The energy sector comprises a single energy provider owned by the government (and Shanghai Electric Power) and service industries (such as car importers or tourist establishments) benefit from government 'green' programmes as well as subsidies. Meanwhile, as a result of government schemes, citizens also have access to 'green incentives', from support for installing solar panels to free bags for the disposal of recyclable waste or home packs to minimise water consumption.

This government drive towards climate neutrality is widely supported by the public, with 92% of the Maltese population defining climate change as a 'very serious' problem in 2019 (against an EU average of 79%), while 97% supported the EU's aim of becoming climate-neutral by 2050 (against an EU average of 92%). However, even though government aims align with those of the public, Malta's climate neutrality aspirations are looking increasingly fragile.

The European Union and Malta's 'Lack of Ambition'

While Malta has committed to lowering its emissions by 19% over the 2021–2030 period (under the Effort-Sharing Regulation), the Commission actually expects Malta's domestic emissions to increase by 41% with the country aiming to meet its reduction targets by transferring annual 'credits' from other Member States. Malta has made it clear that consistent growth in the economy, the influx of workers, increased urbanisation and high-rise dwellings as well as territorial limitations which restrict afforestation means that Malta cannot achieve further greenhouse gas reductions due to the reality of it being a small, densely populated island and not because of a lack of political will.

Nevertheless, the Commission considered Malta's National Energy and Climate Plan (2019) to 'lack ambition'. In other words, the Commission did not feel that the government's arguments based on mitigating circumstances sufficiently explained its inability to meet future targets and hence recommended more ambitious thinking to address Malta's small, island status. In response, the government has stated that the carbon neutral goal continues to be a main pillar of the economy going forward and promised a number of new initiatives, including pilot projects to develop floating wind and solar farms off the country's coast. Ultimately, Malta is not a major polluter and has made marked progress in lowering emissions over the last decade, hence climate change measures are unlikely to become overly demanding in the near future as they are not a major determinant of Malta-EU relations.

Recommendations

Going forward, more EU-funded data and expertise ultimately benefits small countries such as Malta where technical 'know-how' is limited. That said, a transformation in mind-set is needed. As a small island state, government and public thinking always begin with the statement that Malta (small, peripheral, densely populated) is exceptional, unique and therefore eligible for exemptions, thereby keeping public policy firmly within a box. Being heavily centralised under a single-party government with a perspective based on seeking exemptions, Maltese (climate change) planning needs more public-private interface, more decentralised thinking and less focus on limitations. Beyond Malta, the EU needs to continue focusing on 'green' funding as well as investing in research and technology because consumer-driven economies cannot expect their citizens to mainstream 'green practices' if the cost significantly undermines disposable income, no matter how stark the warnings might be of climate catastrophe.

Mark Harwood is an Associate Professor at the University of Malta and a Director of the Institute for European Studies. Having previously worked for the European Commission as well as the Maltese Government, his area of research is the impact of EU membership on Malta.

The Institute was founded in 1991 as a teaching and research institute within the University of Malta. Offering a full range of degree programmes up to PhD level, the Institute has over 1000 alumni. The Institute is also a member of TEPSA.

Concerned But Not Fully Dedicated: The Polish Perspective on Climate Change

Joanna Dyduch, Magdalena Góra, and Natasza Styczyńska

Over the past few years, public debate in Poland has been centred on domestic political matters and the COVID-19 pandemic. However, of late not unsurprisingly the dominant topic has become the war in Ukraine and the resultant economic turmoil. Notwithstanding such major events, environmental issues continue to be of significant importance for Polish society, albeit always a source of disagreement between Poland's government and the EU. According to an opinion poll conducted in 2020, 76% of Poles declare that climate change brings more harm than good, although it seems that people are generally not yet aware of the influence these changes can have on their personal situations. For instance, only 8% worry about water deficits and 4% about problems caused for agriculture. Eurobarometer 2021 demonstrates that only 30% of Poles feel personally responsible for climate change, in contrast to the EU average of 41%. This stance might result from a lack of knowledge about the process of climate change, but also a shallow interest linked to poor education in this matter. Some 75% of Poles want to increase the 2030 national climate target, preferring pro-climate regulations for transport and construction rather than emissions taxes that would be costly for consumers.

Climate change has become increasingly important especially for young Poles living in the various Polish metropolises. However, the overall concern over climate change is still less among Poles than it is amongst Western European societies. Significant impact on raising awareness in regard to the importance of combating climate change has been achieved by urban civil society organisations, particularly in Southern Poland where people are experiencing severe air pollution problems connected with fossil fuels' role in heating systems and energy production. Such organisations as Smog Alarm established in Kraków have made unprecedented efforts and are bringing about change by pushing cities into the vanguard of

J. Dyduch (✉) · M. Góra · N. Styczyńska
Jagellonian University, Kraków, Poland
e-mail: joanna.dyduch@uj.edu.pl; mm.gora@uj.edu.pl; natasza.styczynska@uj.edu.pl

legislation limiting emissions. They have not only successfully lobbied for systemic change, but also mobilised society and raised awareness generally. Interestingly and importantly, there is a section of society that demands more activity from the government and institutions. Serving as a good example is the 2021 precedent which was set when citizens successfully sued the Polish authorities for inaction on the climate crisis. Environmental organisations established 'the Climate Coalition' which not only criticises the Polish government's climate policy, but also proposes expertise and solutions. According to them, the current policy leads both to Poland's further dependence on coal and moreover a threat of its isolation in the world.

The EU's overall environmental policy has always been a source of disagreement between the Polish governments and EU institutions. Even during membership negotiations, it was claimed that legislation in the chapter 'Environment Policy' constituted one of the most expensive areas where Poland had to adapt on a domestic level, very often involving massive investment in infrastructure. It is also significant that Poland received its longest transition periods for adopting some aspects of environment protection. Since 2004 when the policy legislation and high standards of environment protection became binding in Poland, a number of significant conflicts emerged between successive Polish governments and the European Commission. An example is the Rospuda case in which Poland violated the Nature 2000 directives so as to fast-track highway construction in 2006/2007. These events solidified the country's position as an environment protection sceptic. This has been heightened whenever the right-wing Eurosceptic Prawo i Sprawiedliwość has been in power (in 2005–2007 and since 2015 onwards), but more centrist Platforma Obywatelska governments (2007–2015) have fared little better. Overall, in Polish political discourse, the key line of conflict is to what extent can the government allow environmental concerns to restrict development goals and sovereign decisions in that regard (the latter particularly for right-wing parties). Climate policy has a significant social dimension in which the costs of transformation may disproportionally affect low-income households and are also very often—as in the case of the Polish coal region Śląsk—territorially unevenly distributed. Similarly, to many other Central and Eastern Europe countries, this has fuelled domestic narratives which typically blame the EU for these additional costs. Overall, such factors have resulted in the European Commission (and the EU as a whole) being positioned on the opposing side of a political conflict in Poland.

Following the European Commission's 2018 strategy of 'Clean Planet for All' and the European Council's 2030 target for reduction of net greenhouse gas emissions by at least 55%, the EU is ultimately aiming at achieving climate neutrality by 2050. However, some Central and Eastern European countries, Poland in particular, are facing difficulties in meeting these ambitious environmental goals since they still rely very heavily on fossil fuels, while renewables and biofuels are less prominent in their energy mixes compared to the EU average. Accordingly, the Polish Government throughout negotiations has been signalling that the country's energy ecosystem is not yet ready to implement drastic reforms in the time frame proposed by the Commission, arguing that Poland's energy transition requires more time and financial resources. 'Polish Energy Policy by 2040' assumes that Poland would be ready to obtain energy neutrality only in 2050. However, the country has yet to achieve full implementation of the EU's energy packages. From the EU Recovery Fund, Poland is entitled to EUR 14.3 billion for green energy transition, a key factor within 'The National Reconstruction Plan'. Among the flagship projects are investments in energy efficiency (by systemic exchange of heat source in residential buildings and the local community facilities), as well as increases in the use of renewable energy sources and development of technologies based on hydrogen along with other decarbonised gases.

It can be argued that the major driver of Poland's green energy transition is pressure from the EU rather than substantial internally rooted environmental considerations. The Polish energy sector to a great extent still relies on coal which has a 66% share in the country's energy mix, with the balance made up of renewable energy sources 25%, natural gas 6% and pumped-storage power plants 3%. Moreover, around 100,000 people are still employed in the 'coal industry' (power plants and mines), constituting a considerable interest group. Hence, it is not surprising that a green energy revolution is often perceived in Poland as being 'forced by Brussels' and among some interest groups this evokes fear, being perceived more as a threat than an opportunity. Moreover, ongoing Russian aggression in Ukraine is impacting Poland's energy mix, the country's approach to energy security and the overall environmental agenda. Already by May 2022, analysts were pointing to the consequent considerable increase of coal in energy production. Following discussion in mainstream disputes on the need for revising Poland's energy security strategy, the key ideas are limiting the role of gas as a 'transition fuel' in the transformation period, using modern coal technologies, investing in nuclear energy and expanding network connections on the North-South, Baltic liquefied natural gas line, thereby facilitating further diversification of energy resources imports.

Not unsurprisingly, among the main opponents of a fundamental energy transition are trade unions representing the interests of coal miners and employees of traditional energy industries branches. There is no doubt that transition toward green energy will generate costs, both financial as well as political and social, which of course are harder to estimate. Because of such profound transformations in the fuel and energy sectors, energy costs are increasing in 2022 not only for the entrepreneur, but also for individual consumers, hitting the household sector particularly severely. The National Bank of Poland forecasts released in July 2022, state that the average increase in total energy prices will reach 33.9% in 2022 and 30.9% in 2023. If these forecasts turn out to be accurate, then of course this will have a tremendously detrimental impact on Poland's economic situation.

Nevertheless, looking more positively the share of renewables in electricity production is increasing and reached 17.9% in 2020, whilst the share of renewables in gross final energy consumption is, in turn, slightly over 16.13%. Moreover, for the first time the total share of coal in the country's energy mix was lower than 70%. In 2009, this level was 90% and in the early 1990s, even 96%.

Among the champions of a green transformation are mostly state-owned giants. In 2020 it was the ORLEN Group (oil refiner and petrol retailer) and then in 2021

Polish Energy Group took the lead. ORLEN is investing in solar farms, biogas plants and hydrogen production, as well as developing a network of electric car chargers and producing green glycol. Meanwhile, the Polish Energy Group is working in partnership with the Danish company Ørsted, constructing the first large-scale offshore wind farm in the Baltic Sea. Recently it has also invested in the development of prosumer photovoltaics and energy storage. Other Polish energy companies such as TAURON, ENEA, PGNiG, ENERGA, PGNiG and KGHM are also investing heavily in removable technologies and infrastructure.

Recommendations

Recent turmoil in regional and global energy markets, resulting from Russia's invasion of Ukraine, has sparked immediate reactions by the EU and its Member States, thereby accelerating steps aimed at decreasing and eventually eliminating any dependency that European markets may have on Russian fossil energy resources. As early as March 2022, the Polish government announced a cessation in the purchase of Russian coal and will now aim to stop importing oil and gas from Russia by the end of 2022.

While designing wide-ranging responses to the current geopolitical and geoeconomics re-evaluations, attention should be paid not only to seeking new directions for energy resources supplies, but even more to accelerating development of the comprehensive electro-energetic system based to a far greater extent on renewables. Here the need for further well-designed investments in critical infrastructure (electricity transmission infrastructure, strengthening the domestic transmission of energy resource system and storage capabilities) is particularly important.

Although according to the 2021 Eurobarometer 90% of Poles support the reduction of greenhouse gas emissions and the pursuit of climate neutrality in the EU by 2050, knowledge about climate change remains very vague with a perception of some distant problem. Polish society needs more information and educational campaigns raising environmental awareness but also an explanation of what can be done on the micro/local level. This could be undertaken not only by the state but also by grassroots organisations that are more trusted by citizens and hence may have a bigger impact at local levels.

Joanna Dyduch is Associate Professor at the Institute of Middle and Far East of the Jagiellonian University in Krakow, She is a political scientist and international relations scholar. Her research interests lie in the area of energy studies (mostly Israel and European). Joanna is the author of books and articles on energy policy, securitisation and de-securitisation of energy issues, Europeanisation and de-Europeanisation of member states energy policy (especially in its external dimension). Additionally, her research activities touch upon intersections between the foreign policy and energy-related issues.

Magdalena Góra is an Associate Professor of political science and European studies at the Institute of European Studies of the Jagiellonian University. Her research deals with legitimacy and contestation in external relations of the European Union, EU actorness in international relations, especially in EU's close neighbourhood, as well as democratic challenges in the EU. She has worked and taught in a number of academic institutions worldwide as well as having published several peer-reviewed journal articles, book chapters and co-edited volumes.

The Institute for European Studies is part of the Faculty of International and Political Studies at Jagiellonian University—the oldest and leading university in Poland. The Institute is famous for its interdisciplinary approach that combines the perspectives of anthropology, economy, cultural studies, political sciences, history, law and sociology. It is also an Associate Member of TEPSA.

Natasza Styczyńska is an assistant professor at the Institute of European Studies of the Jagiellonian University. Currently, she is a researcher in two H2020 projects: Populist rebellion against modernity in twenty-first century Eastern Europe: as well as neo-traditionalism and neo-feudalism (POPREBEL) together with EU Differentiation, Dominance and Democracy (EU3D). Her academic interests include transformation processes in Central and Eastern Europe, party politics, nationalism, populism and Euroscepticism in the CEE region and the Balkans.

Portugal Facing Climate Change: Deep Problems, Sluggish Responses, But Hopeful Prospects

Luisa Schmidt

In Portugal, the importance of fighting climate change is clearly acknowledged, as an expression both of public opinion and official recognition. Public interest stems largely from three key sources: a persisting rustic sensibility; the deepening role of environmental education, notably in schools; soaring media coverage. Official recognition manifested in legislation, institution mandates and supportive public policies, has emerged only over the past 15 years, with government investment in renewable energy and the publication of national plans to fight climate change, which have since then been strengthened.

Public perceptions and associated policies to deal with climate change have greatly improved in the aftermath of national catastrophes in 2014 and 2017. In January and February of 2014, a number of storms, especially Hercules, wreaked havoc on the Portuguese Atlantic Coast causing significant damage to ports, marinas and beach equipment, as well as destroying dunes and removing sand from beaches. The total cost of subsequent repairs amounted to EUR 23 million. Even worse and much more dramatic were the wildfires of 2017, which burned 540,000 ha of forest, shrubland and agricultural areas. As a result, 66 people died and over 200 were injured. In the aftermath of this tragedy, the Agency for the Integrated Management of Rural Fire was created, which reports directly to the Prime Minister. However, successive events have caused setbacks to climate change-related actions, such as the financial crisis and challenging austerity period thereafter, the COVID-19 pandemic and the current upheaval caused by Russia's invasion of Ukraine.

The country's pathway towards climate neutrality is strongly hampered by factors pertaining to geographic vulnerability, poorly managed spatial planning and high dependency on external energy sources. Regarding this third factor, the high level (65.8% in 2020) is due to the country's primary energy mix, which largely comprises

L. Schmidt (✉)
Institute of Social Sciences - University of Lisbon, Lisboa, Portugal
e-mail: mlschmidt@ics.ulisboa.pt

© The Author(s), under exclusive license to Springer Nature Switzerland AG 2023
M. Kaeding et al. (eds.), *Climate Change and the Future of Europe*, The Future of
Europe, https://doi.org/10.1007/978-3-031-23328-9_21

oil (40.9%), mostly from Brazil; natural gas (25%), mainly from Algeria; coal (2.7%) and others (1.6%). Only 29.9% of this mix results from domestic renewable sources (predominantly hydropower and wind power).

Portugal is particularly prone to climate change impacts, notably: extreme sea-level changes and severe marine events; desertification, drought and water scarcity; as well as heatwaves and wildfires. Climate policies in Portugal are heavily influenced by fossil fuel-based energy dependency and ambivalent strategies to tackle this. Policy tools such as the National Energy and Climate Plan (PNEC 2030) and the Roadmap for Carbon Neutrality 2050 are currently in place. These not only encourage decarbonisation policies but have also propelled a significant expansion of renewable energy investments and accelerated the recent closure of coal-fired power plants. Greenhouse gas emissions have thus decreased, albeit with fluctuations notably in the past five years.

However, there remain awkward structural obstacles to achieving full carbon neutrality. Particularly relevant here is the dominance of road transport (for both passengers and goods), together with poorly designed spatial planning and energy-inefficient housing stocks. The growth of outer suburbs around big cities was unfortunately not accompanied by carbon-reducing mobility improvements. Portugal has been experiencing the lowest rate of public rail transport usage in the EU and is only now starting to pick up thanks to much needed investment. Meanwhile, energy requirements in ill-designed buildings continue to account for 30% of the country's final energy consumption.

In a country where the construction industry plays a major role in the economy and employment, achieving climate neutrality also entails the reconversion of heavy industries with high energy intensity and significant carbon footprints, such as cement plants. The same applies to ceramic, glass and petrochemical industries. The industrial sectors with the highest levels of total emissions are cement (20.06%) and petrochemicals (16.6%). By contrast, over the past 15 years, Portugal has witnessed development and growth in the renewable energy sector, which is ultimately contributing to greater energy autonomy for the country and consequent decarbonisation. This sector now covers 58% of electricity generation and is expected to reach 80% by 2026.

The country suffers regularly from large wildfires, due to rural abandonment and forest mismanagement. Indigenous forested areas have decreased, fire-prone non--native trees have been planted and combustible fuel is widespread. In turn, the mass tourism sector, which is heavily reliant on real estate as well as transportation and moreover encourages spatial mismanagement, hampers the achievement of climate neutrality goals.

Amongst positive strategies to combat climate change is a network of adaptive municipalities (adapt.local 2017). This network has focused on carbon-reducing planning by creating the right conditions for ensuring smooth mobility, establishing green spaces for public health and heat-stress mitigation in cities, as well as encouraging the installation of solar panels.

Poorly managed public administration, flawed bureaucratic procedures and a civil society traditionally distanced from the state provide powerful obstacles to the effectiveness of climate change policies. Increasing public communication and expanding environmental education, as well as creating mediating agencies between citizens and public administration, would help to overcome these frictions.

Portuguese people express increasing concern over climate change. In both the last 2018 European Social Survey and 2021 Eurobarometer on the Environment, they appear as 'most concerned' with climate change and are amongst those who value renewable energies (wind and solar) very highly. However, in a country where 22% of people live at risk of poverty, where the average salary is low and where housing costs have increased tremendously, it is socioeconomic survival which has become a priority. Despite its mild weather, Portugal ranks third in the EU for energy poverty, due to rising prices and inefficient housing. The social dimensions of energy transition in Portugal are thus highly demanding in political terms, which has led to the enactment of compensatory measures. Examples include financial support for electricity bills and direct subsidy for household energy efficiency conversion. However, these measures do not tackle any social challenges associated with green transition from a structural perspective, which require more citizen-centred public investments and energy policies.

Support provided by European funds has been particularly important in overcoming some of these problems, notably those related to energy transition— the democratisation of renewable energies, thermal comfort of buildings and railway development. European funds underpinned approval of the post-COVID-19 Recovery and Resilience Plan (2021–26), which devotes 37% of its total allocation (EUR 16.644 million) to climate change.

In the European context, Portugal is well positioned to increase its production of green energy, either through renewables (solar and wind offshore) or through green hydrogen. Moreover, it has the third largest Maritime Exclusive Economic Zone in the EU, which is subject to enlargement. This Zone is designed to safeguard Portuguese territorial waters, with the sea being recognised as a notable carbon sink in providing important fishery and bio-marine resources for delivering the European Green Deal. Within the EU, Portugal is a biodiversity hotspot and as such can seize opportunities to restore, manage and invest in its Protected Areas, nature conservation as well as the creation of sustainable forests. Concerning agriculture, the country could provide an important contribution to the 'Farm to Fork' strategy as it holds optimal conditions for proximity farming—organic, integrated and local. Notwithstanding these positive aspects, Portugal is nevertheless still partly embarking upon intensive and unsustainable agricultural production systems.

Recommendations

To improve its contribution to European cohesion for climate action, Portugal must seek strong European support on four topics, which stimulate recommendations for further governmental measures.

Firstly, energy conversion of transportation has to be undertaken with special emphasis on power and rail connectivity with the EU, which will mean a high potential for reducing CO2 emissions. Likewise, provision of energy transport connectivity between the Iberian Peninsula and the rest of Europe is essential, both in terms of the gas pipeline and renewable electricity.

Secondly, water policies should be integrated across Iberian hydrographic basins. Greater attention and more support must be given in countering the extreme drought situation that the country is currently experiencing, a situation that is likely to worsen in the coming years due to climate change.

Thirdly, a reformed Common Agricultural Policy must be introduced to fight climate change and protect biodiversity, which converges much more closely with the European Green Deal policy 'Farm to Fork' and contributes to its organic farming goals.

Finally, the reinforcement of environmental education must be established at all levels of education and civil society organisations, as must integrate municipal and intermunicipal policies as well as strategies to mitigate and adapt to climate change.

Coping with current inflationary pressures and energy/security issues in the aftermath of the Ukraine war must be integrated with both mitigation and adaptation investment on a European scale. Portugal is part of the EU's emerging strategy for producing clean energy and collective energy security as well as tackling energy-related social poverty. Given the country's current levels of fossil fuel-based energy dependency, impacts on the availability and prices of imported primary energy sources are foreseeable, even indirectly. Moreover, impacts on land uses are expected, because of the need to increase food security, as well as a growing demand for biofuel at a time when the country is in the midst of a severe drought. Portugal is making some progress, albeit launching its net zero future from a high level of structural carbon dependency.

Luisa Schmidt is a sociologist and Senior Research Fellow at the Institute of Social Sciences of the University of Lisbon (ICS-ULisboa). In Portugal, she pioneered Environmental Sociology research and outreach. Amongst her achievements, she: co-founded and leads OBSERVA—Observatory of Environment, Territory and Society; developed the first large survey about environmental values and practices (2000); put together the first large longitudinal environmental media analysis (2009); and the international public consultation on 'Climate and Energy' COP21 Paris (2015). She co-founded the interdisciplinary PhD on 'Climate Change and Sustainable Development Policies' being part of its scientific committee. She currently runs and/or integrates several projects on this field of work, having written 20 books and over 120 scientific articles. ICS-ULisboa is a top research Institute on Social Sciences and is part of the University of Lisbon (ULisboa), which resulted from a merger of the former University of Lisbon and the Technical University of Lisbon. Currently, ULisboa is the largest University in Portugal and one of the largest in Europe. Located in the heart of the city, it presents its entire community one of the best and most diversified training offers.

Romania's Fight Against Climate Change. Contributing to Ambitious European Targets While Facing Deep-Rooted Sectoral Flaws

Eliza Vaş and Mihai Sebe

The path towards climate neutrality was institutionalised for EU Member States with the adoption of the European Green Deal. The subsequent pieces of legislation create (d) the legally binding framework for each Member State to achieve the EU's intermediary (2030) and final (2050) targets. Through the 2021–2030 Integrated National Energy and Climate Plan and following the European Commission's recommendations, Romania has committed to the following national objectives by 2030: reduce the Emissions Trading System emissions (−43.9% compared to 2005); reduce non-Emissions Trading System emissions (−2% compared to 2005); increase the share of renewable energy in gross final energy consumption (30.7%); ensure energy efficiency, by achieving energy savings of 45.1% in primary consumption compared to 2007.

As for funds allocated through the NextGenerationEU, Romania's recovery and resilience plan was approved with EUR 14.2 billion in grants and EUR 14.9 billion in loans, with 41% of the reforms and investments proposed through this plan dedicated to supporting green transition. Some domains being funded include water management, afforestation and biodiversity protection, waste management, sustainable transport, building renovations and energy.

The European objectives of reducing greenhouse gas emissions by at least 55% by 2030 and achieving climate neutrality by 2050 have been considered feasible by Romania, so long as energy mix decisions continue to be taken at national level. The national energy mix for the first five months of 2022 comprised: crude oil (32.03%), natural gas (28.53%), coal (10.60%), aggregated hydropower, nuclear, solar, wind energy and imported electricity (17.10%) and imported oil products (10.01%).

Given the ongoing 2022 Russian war in Ukraine and changes in the volume of fossil fuel imports, Romania has used its energy mix prerequisite to update

E. Vaş · M. Sebe (✉)
European Institute of Romania, Bucharest, Romania
e-mail: eliza.vas@ier.gov.ro; mihai.sebe@ier.gov.ro

decarbonisation targets by increasing the use of coal for heat and electricity until alternative sources are operationalised. While a number of carbon-intensive industries are functioning and the geopolitical challenges requires the use of existing energy capabilities, the positive decarbonisation track record that led to reducing carbon intensity per unit of GDP by 54% (between 2005 and 2019) is certainly worthy of acknowledgement.

Romania's fight against climate change foresees using natural gas and nuclear energy as transition fuels. In this sense, a number of developments in the energy sector have been projected for the following decade: exploiting natural gas reserves from the Black Sea, implementing the decarbonisation plan for Complexul Energetic Oltenia (the leading producer of electric energy from coal and lignite), diversifying uranium sources for Nuclearelectrica (the only national producer of nuclear energy) and building new nuclear capacity; increasing renewable energy capacities and interconnecting with the regional market; as well as developing the national electricity grid and the gas infrastructure. In addition to this, Romania has become more interested in building its hydrogen capacity both for industrial purposes and individual consumers. One proposal in the national plan for energy and climate change foresees analysing the possible injection of hydrogen into gas installations as part of the country's decarbonisation roadmap.

Besides Romania's national reserves, there is another key reason for which natural gas is considered essential in keeping with the transition process' green taxonomy. Almost half Romanian households use wood for heating, most being in the rural areas where alternative heating sources are lacking. In this sense, local authorities have sent around 1200 project applications to receive funding within the national investment plan either to modernise existing or create new gas infrastructure over the next few years. Investing in gas while the general trend is to decrease the use of fossil fuels can seem counterintuitive. At the same time, public funds to support the installation of solar panels or heating pumps are limited, thus making it difficult for individual consumers to access financial resources to update their heating systems.

Another dimension in Romania's energy outlook is that of a high energy poverty rate. Affecting more than one third of households, energy poverty is both an indicator of citizens' current economic status as well as a sign of future needs to invest in renewable energy and ensure higher levels of energy efficiency. Connected with a lack of green competences, this can compromise the fight against climate change and the implementation of large-scale projects in the long run.

In terms of switching to renewables for producing electricity, Romania has started to adopt funding schemes that allow individual consumers to apply for the installation of photovoltaic panels. However, limitations are readily apparent in: the relatively low level of funding and the need to advance total sums that will be reimbursed at some later stage (the waiting time often amounts to two years); the lack of know-how (understanding the changes that need to be implemented and the different options available on the market, as well as finding a professional official contractor); and missing green competences, namely those required for transitioning from consumer to prosumer.

In the fight against climate change, the green transition of the transport sector must also be tackled. Domestic transport ranks third in the sectoral shares of greenhouse gas emissions, with 23.80% recorded in 2020. Here, Romania displays major vulnerabilities regarding its vehicle fleet (one of the oldest in the EU). Over the past two decades, second-hand vehicles from Western Europe have been subject to widespread importing into Romania, thus making worse an already heavily polluting car fleet, given the usage of petrol and diesel. An opportunity deriving from domestic capacities to produce electric energy could give rise to an accelerated transition to electric cars and increase in the coverage of charging stations foreseen for the next few years.

Another challenge related to climate change this time at EU level is the need to ensure fair conditions and implementation, taking into account Member States' specificities, including Romania. It should be ensured that transition to climate neutrality is both socially sustainable and in line with each Member's economic development, rather than extending still further different levels of development.

Considering the particularities that characterise the path towards a low emission economy and climate neutrality, it is of no surprise that the Eurobarometer results from March to April 2021 indicate that less than 10% of respondents from Romania consider climate change as the most important issue the world is facing. Challenges related to the spread of infectious diseases, poverty and the economic situation are seen as more pressing, while climate change and the deterioration of nature come next on the list of priorities. In terms of acting on climate change, the same survey shows that Romanian citizens look up to the EU first, and the national government second.

Recommendations

These particularities suggest various courses of action, enabling national and European decision-makers to act steadily on climate change.

Firstly, national decision-makers must increase levels of knowledge and develop green competences among Romanian citizens, providing them with a better understanding of how climate change affects both households and the general environment. The social and economic costs of adapting to the new reality must be made clear. The energy market's liberalisation process has generated complex issues for citizens, such as bills at unpredictable levels, misunderstandings regarding the terms and conditions of new contracts and the benefits of a competitive market. All these issues present opportunities for educating the public ready for similar processes in future.

Secondly, create large-scale projects (that can be implemented by local authorities) with the purpose of renovating and modernising households to derive further implementation of renewable energy solutions (solar thermal panels for heating and photovoltaic panels for electricity). Provide transparent and easy-to-access information on how households can apply for such funding and offer free guidance to citizens that require assistance in developing an application.

Finally, European decision-makers must support tailored approaches for citizens and regions that need more time and investments not only to adapt to new policies, but also to update their current models of living (especially those residing in rural areas). They should also develop and combine the network of green diplomatic missions and consolidate climate diplomacy in order to ensure both European and worldwide support.

Eliza Vaş is an expert within the European Institute of Romania, PhD candidate in International Relations and European Studies at the Babeş-Bolyai University as well as Policy & Strategy director within the Young Initiative Association. She has a master's degree in Comparative European Political Studies, awarded by the Babeş-Bolyai University and Université Paris-Est Marne-la-Vallée. She has extensive experience in the non-profit sector, and has (co)authored articles about democracy, civil society, civic participation, European Green Deal, circular economy, youth policies and new technologies.

The European Institute of Romania (EIR) is a public institution whose mission is to provide expertise in the field of European Affairs to the public administration, the business community, the social partners and the civil society. EIR is also a member of TEPSA.

Mihai Sebe is an expert in European affairs and Romanian politics, currently coordinating the European Studies Unit at the European Institute of Romania and guest instructor at the Faculty of Political Sciences, University of Bucharest. He holds a PhD in Political Science from the University of Bucharest. His area of expertise includes topics such as the history of the European idea, populism, future politics. He writes extensively on European politics both at home and abroad.

The European Institute of Romania (EIR) is a public institution whose mission is to provide expertise in the field of European Affairs to the public administration, the business community, the social partners and the civil society. EIR is also a member of TEPSA.

Climate Change: A Second-Class Agenda in Slovakia?

Donald Wertlen

The Climate Agenda in Slovak Politics

The Slovak Republic has been a ratifying party to the Kyoto protocol, the Paris Agreement and as a Member State of the EU, has committed to delivering carbon-net neutrality by 2050. However, despite climate change being an omnipresent agenda in Slovak politics, it has never established itself as a dominant political discourse, often being side-lined by other pressing issues—most recently (since 2015) the migration 'crisis', COVID-19 pandemic or war in Ukraine.

Regarding its own initiatives on climate change mitigation, the Slovak government is (and has usually been) extremely passive. Most activities, such as increasing the share of renewable energy or improvements in energy efficiency, are a result of 'downloading' the EU-inducted policies and measures. In this sense, if it were not for EU membership, the situation regarding Slovakia's climate change responses would be significantly worse. Hence, immediate prospects for Slovakia's green transition, which is driven by EU policies, are very much dependent on the successful 'Europeanisation' of its climate (and energy) policy. The major problem is a lack of political motivation to push the climate agenda forward. The green parties are virtually non-existent and scarcely active in elections. The Green Party (Strana zelených), which was a member of the European Greens, has achieved less than 1% of the votes in national elections (2002 and 2012) and less than 0.5% of votes in European Parliament elections (2014). Moreover, the Slovak Green Party (Strana zelených Slovenska), which fragmented from the Green Party (Strana Zelených), has had similarly abysmal results gaining less than 0.8% in both national (2016) and European (2019) elections. Indeed, the climate agenda appears very much on the periphery of Slovakia's political mainstream spectrum of programmes. In any event,

D. Wertlen (✉)
Comenius University Bratislava, Bratislava, Slovakia
e-mail: donald.wertlen@fses.uniba.sk

99

having a green agenda in Slovakia neither wins you political points nor resonates among voters sufficiently to form the cornerstone of a political campaign. The only relevant political stakeholder that actively promotes the climate agenda is the Slovak Republic's President, Zuzana Čaputová, albeit her competences are limited as the legislative and executive powers belong to parliament and government. Despite that, the President still has agenda-setting abilities and can make symbolic gestures—for example the 'Green Office' initiative by which Presidential Office aspires to become the first climate neutral public institution in Slovakia by 2030.

Public opinion is quite difficult to gauge, given Slovakians' peculiar climate change bias. In other words, when targeting the public's perception of climate change and its consequences, most agree that it is 'very serious problem'. However, when compared with other issues in more generally oriented surveys, climate change concerns often fall behind others, such as inadequate health and education or the fight against corruption. Hence, the lack of demand for stringent climate policies coupled with the absence of any direction from central government have resulted in the Slovakian agenda's dormancy. However, on a more positive note, the 'bottom-up' approach has of late been gaining momentum with the civic initiative 'Klíma ťa potrebuje' (Climate Needs You), which requests the government not only to declare a state of climate emergency, but also stipulate a complex framework strategy for climate neutrality in Slovakia. The biggest online petition in Slovakia's history has been signed by almost 130,000 citizens and made its way into the National Council. Unfortunately, besides the petition's declaratory acknowledgment by the National Council, no tangible outcomes have as yet been made, which consequently has resulted in the petition's ongoing second round.

Challenges

Regrettably, the war in Ukraine has failed to act as a catalyst for any 'greening' of Slovakia's energy mix. In light of the increasing electricity prices, energy security has effectively 'stolen' the narrative, particularly regarding affordability with Russian gas still presenting a viable option. As the Slovak republic has almost 90% natural gas and 100% oil import dependency (from which 85% of gas and all oil come from Russia), any diversification strategy is preoccupied with suppliers rather than sources. Moreover, continual advocacy for nuclear energy as green or sustainable fuel prevents Slovakia from taking a momentous leap towards renewable energy.

One of the major challenges in the country's climate agenda is decarbonisation of the Slovak economy, particularly the heavy (steel) industry. US Steel in the Košice region is the largest polluter (together with its subsidiary Ferroenergy), producing more than 40% of Slovakia's CO2 emissions within the EU-wide emissions trading system. Hence, green transformation of the steel sector is imperative if there is going to be a successful reduction of greenhouse gases in Slovakia and this requires massive investments in the electrification or potential utilisation of hydrogen in production processes. By contrast, decarbonisation does offer an opportunity to spearhead a 'green steel' market, which would play a significant role in the future of energy and transport sectors.

Furthermore, Slovakia is one of the EU Member States with identified coal regions, specifically, Horná Nitra. Even though there is a gradual declining trend, coal mining is still the region's economic backbone and biggest employer. In this sense, Slovakia must advance the coal phase-out strategy (both mining and power plants), as this is the most carbon-intensive fossil fuel. At the same time, central government has to ensure a just transition, mitigating negative impacts on the region's employment and energy supply. Moreover, one of the key issues is the environment's rehabilitation, given its severe degrading by long-term exposure to mining activities.

To overcome these (and other) challenges, Slovakia has allocated more than EUR 2 billion (approximately one-third) from the EU's post-pandemic financial package 'Next Generation EU' for the 'green economy' within its national recovery plan. Although the policies and legislative measures are formulated, we have yet to see how successful their implementation will be.

Recommendations

Firstly and of key importance, the Slovak government must articulate a comprehensive and concrete climate plan, with a focus on the integration of climate mainstreaming into all policy sectors. While utilising available finances from the EU (such as Next Generation EU), the government must take proactive steps to engage municipalities in the formulation and implementation of clear but contextual climate policy objectives.

Secondly, the government should mobilise investment in environmental research and innovations, as Slovakia is currently among the worst-performing EU Member States in this regard. This would also help in a domestically driven decarbonisation of the economy, heavy industry and carbon-intensive regions.

Thirdly and finally, the war in Ukraine has provided an impetus for diversification from Russian energy supplies (which are dominant in the Slovakia's energy mix, constituting almost 60% of total energy imports). Although it opens a window of opportunity to shift the energy focus onto renewable sources and energy efficiency, it is crucial to avoid new fossil fuel 'traps' (for instance liquefied natural gas) that could lead to future lock-in on hydrocarbons.

Donald Wertlen holds a PhD degree from the Institute of European Studies and International Relations at Comenius University in Bratislava. Energy and climate policies and differentiation in the European Union have been primary points of his research interest. In his publications, he has deliberated on different aspects of the energy and climate policies interplay, such as energy security, energy poverty, human rights or space policy. Moreover, he has been member of research teams of several national research projects/grants. Currently, he is teaching at the Faculty of the Social and Economic Sciences CU.

The Faculty of Social and Economic Sciences is an integral part of Comenius University in Bratislava. Academics and researchers provide expertise in different fields of social science for national decision-makers, running research and popularisation projects in Slovakia and abroad. The faculty's foreign professors, students from abroad and European research projects give it a truly international feel. In the last decade, it has earned a reputation as one of the best social science faculties in Slovakia. The Institute is also a member of TEPSA.

Slovenia: Big on Plans, Small on Deeds?

Danijel Crnčec

Slovenia has travelled on a long and bumpy road in tackling climate change. More than a decade ago, the issue was already high on the government's list of priorities. In 2009, the Government office for climate change was established. Soon after, the Climate Change Act was drafted and an update of the National energy programme was proposed with ambitious energy targets for 2030. With a change of government in 2012, the office was disbanded, with both the draft act and energy programme being shelved. It was not until 2020 that the first signs of a more ambitious and decisive policy returned with adoption of the integrated national energy and climate plan, followed by the Slovenian climate long-term strategy 2050 (2021) and the national coal phase-out strategy (2022). Finally, the government elected in April 2022 made green transition one of its top priorities and announced a new Ministry of Climate and Energy to lead the country's green transition.

Many Challenges, Few Answers

In 2021, Slovenia decided to become climate neutral by 2050. Two scenarios were developed—one with long-term use of nuclear energy and the other with increased use of renewable energy and synthetic gases. Many challenges on the path to climate neutrality became clear, but few answers were provided.

Transport (passenger and freight) has become heavy and unsustainable, accounting in 2020 for the largest share of final energy consumption (40%), with its share of greenhouse gas emissions having increased by 27% since 2005. Moreover, its emissions in 2030 are projected to be 10% higher than in 2005. Appropriate and

D. Crnčec (✉)

Centre for International Relations - University of Ljubljana, Ljubljana, Slovenia

e-mail: danijel.crncec@fdv.uni-lj.si

© The Author(s), under exclusive license to Springer Nature Switzerland AG 2023

M. Kaeding et al. (eds.), *Climate Change and the Future of Europe*, The Future of Europe, https://doi.org/10.1007/978-3-031-23328-9_24

103

effective climate policies for the transport sector will be critical to reducing green-house gas emissions in the long term.

The country has to resolve a number of issues related to the use of nuclear energy. The existing Krško nuclear power plant, jointly owned by Slovenia and Croatia, originally planned to operate from 1983 to 2023, was to be extended until 2043. However, the administrative procedures (including strategic environmental assessment) have yet to be completed. Also, the two countries have not agreed on a joint radioactive waste storage facility. While each is building its own storage for medium- and low-level radioactive waste, high-level radioactive waste and spent fuel will remain in Krško until 2043. In 2021, the Slovenian government decided to support the long-term use of nuclear energy, without consulting the public, and hence issued an energy permit for a new nuclear power plant in Krško. However, in 2022, the new government (although supportive of nuclear energy) announced that a referendum would be held on the issue.

Slovenia is lagging behind in terms of the planned deployment of renewable energy and has thus set a very unambitious 2030 target. Indeed, in 2020 the country ranked last amongst EU Member States for its share of electricity generated from solar and wind power (1.75%). Although nine out of ten Slovenians believe it is important to set ambitious renewable energy targets (at national and EU level), civil society and non-governmental organisations are extremely hesitant when it comes to the use of wind and hydropower, due to its potential negative environmental impact. There is a significant implementation gap here. In addition, 38% of Slovenia's territory is protected by Natura2000, which represents a no-go area for deployment of renewable energy. Accordingly, the country will be restricted to using mainly solar energy in attempting to achieve a larger share of renewable energy by 2030.

The country will phase out coal by 2033 at the latest, albeit a decision on alternative power generation has yet to be made. Two coal regions have been earmarked as being eligible for support from the EU's Just Transition Fund: (i) Savinjsko-Šaleška—the lignite coal mine in Velenje and the Šoštanj thermal power plant will be closed. The latter is responsible for about one-third of the country's electricity generation (Unit 6 was originally planned to operate from 2014 to 2054); and (ii) Zasavje—the coal mine and thermal power plant have already been closed, but as a result the region has suffered economically and socially. It remains one of the least developed parts of Slovenia.

Energy transition will also have a significant impact on energy-intensive industry in Slovenia, which accounts for almost two-thirds of industry's total energy consumption. It is largely dependent on the use of natural gas, which comes from the Central European Gas Hub in Baumgarten, Austria, but is of Russian origin. Due to the war in Ukraine, the government has called on all industries to increase their energy efficiency and look for alternatives to (Russian) gas, if possible by using renewable energy. While these measures will accelerate energy transition, any potential gas shortage would pose a significant challenge to the country's gas supply security in the short term.

Last but not least, in addition to the extensive financial support required, a lack of human resources in particular has been identified as the biggest obstacle and risk to successful implementation of the planned additional climate and energy policy measures by 2030.

Nimby vs. Banana?

Whilst Slovenians are generally considered to be environmentally conscious, paradoxically the 2021 Eurobarometer survey found that climate change is considered less important in Slovenia than the EU average. Interestingly, Slovenians are less likely to say they are personally responsible for combating climate change, yet they strongly support the adoption of ambitious renewable energy targets. However, the deployment of renewable energy projects (particularly large-scale wind and hydropower) increasingly faces opposition from local communities and environmental non-governmental organisations. From an initial 'Not in my backyard (NIMBY)' effect, this has evolved into 'Build absolutely nothing anywhere near anything (or anyone) (BANANA)' and hence one of the biggest challenges to further renewable energy deployment in the country.

The EU as Carrot and Stick for Slovenia's Climate Policy

Slovenia's integrated national energy and climate plan was finally adopted because it was one of the conditions for accessing EU funds in the 2021–2027 period. Adoption of Slovenia's national coal phase-out strategy enabled Slovenian coal regions to access the EU's Just Transition Fund. Early in 2022, the previous Slovenian government (despite the climate-sceptic leading party and strong opposition from the local population) even adopted the most ambitious coal phase-out scenario (2033 rather than the originally expected 2038) due to the rising price of EU Emissions Trading System allowances. Slovenia's recovery and resilience plan devotes 42% out of EUR 2.5 billion to climate objectives and includes reforms targeting renewable energy, sustainable mobility and energy efficiency.

Slovenia has tried to position itself as an ambitious green EU Member State. However, the last decade has shown that the country faces significant challenges and implementation gaps in some areas that are critical for achieving its ambitious green transition. EU funds under the Recovery and Resilience Facility could provide an important boost, even though the recovery and resilience plan was prepared without expert and public participation. The new government, which took office in June 2022, has made green transition one of its priorities and promised to intensify cooperation with experts and the public.

Recommendations

Firstly, the government needs to launch open and inclusive public consultation to agree on more ambitious national climate and energy targets for 2030, which should be aligned with the higher EU 2030 targets (stemming from the Fit for 55 and REPowerEU legislative packages).

Secondly, a number of comprehensive scenarios should be developed to achieve the proposed higher national 2030 targets, including simplification and speeding up of permitting procedures for renewable energy projects.

Thirdly, the government should systematically improve climate literacy and capacity building at all levels and in all sectors to close the implementation gap of the past decade successfully. Only in this way will higher climate and energy targets by achieved. In parallel, the European Commission should recognise the country's relevant national circumstances by strengthening its technical and financial support for promoting capacity building at all levels and in all relevant sectors.

Danijel Crnčec is an Assistant Professor at the Faculty of Social Sciences at the University of Ljubljana. His main field of teaching and research is international relations and international law, EU energy and climate policy. He also works for the Ministry of Infrastructure, covering in particular preparation of the Integrated National Energy and Climate Plan. The author offers thanks to Professor Maja Bučar (Centre of International Relations) for her valuable comments. This research was financially supported by the Slovenian Research Agency, Project code: P5-0177 (Slovenia and its actors in international relations and European integrations).

The Centre for International Relations conducts interdisciplinary research in the fields of international relations, international economics and international business, politics of international law, diplomacy, human rights, international organisations and European integration. It is also a long-standing and active member of TEPSA, as well as a number of other international networks.

Solving the Spanish Climate Conundrum While Contributing to the EU's Decarbonisation Compass

Lara Lázaro-Touza

Spain is known to be a climate hotspot within the EU. Scientific reports and the latest National Adaptation Plan indicate that the country can expect an exacerbation of pre-existing weather patterns. This implies, *inter alia*, increases in extreme temperatures and reductions in mean precipitation by up to 30% in the summer, affecting not only the availability and quality of water, but also impacting key economic sectors such as tourism. Conscious of the need to insure against the climate crisis, whilst at the same time being mindful of the economic opportunities and the costs of low-carbon transition, the EU's climate neutrality transition arsenal (the European Green Deal, the Fit for 55, Next Generation EU and REPowerEU) is being supported by the current Spanish government and downloaded to Spain. However, it is receiving a mixed reception from political parties, the business sector and citizens.

According to the Joint Research Centre's PESETA IV report, climate change will have significant and asymmetric economic impacts on selected sectors in the EU. However, as yet, data and evaluations are limited and hence underestimate the total economic impact of climate change, with Southern countries, including Spain, being more severely affected. Analyses of the economic impact that a mean temperature increase of 3 °C would have on current economies, a temperature rise compatible with existing climate action, show that Southern EU gross domestic product could fall by 2.7% in the seven sectors analysed (almost doubling the EU average), albeit other authors estimate substantially larger reductions. Mortality figures contribute the most to estimated economic impacts, followed by droughts, coastal floods, losses in agriculture, river floods and energy.

L. Lázaro-Touza (✉)
Energy and Climate Programme, Elcano Royal Institute, Madrid, Spain

Business Administration, Centro de Enseñanza Superior Cardenal Cisneros (attached to Universidad Complutense de Madrid), Madrid, Spain
e-mail: llazaro@rielcano.org

© The Author(s), under exclusive license to Springer Nature Switzerland AG 2023
M. Kaeding et al. (eds.), *Climate Change and the Future of Europe*, The Future of Europe, https://doi.org/10.1007/978-3-031-23328-9_25

The expected environmental and economic impacts of climate change are power-ful reasons for Spain to engage actively in climate mitigation and adaptation actions. These reasons are reinforced by citizens' concerns about climate change. A survey conducted by the Elcano Royal Institute in 2019, ahead of hosting the United Nations Climate Conference COP25, revealed that Spaniards believe climate change is the greatest threat to the world, followed by armed conflicts, with only 3% of interviewees stating that climate change does not exist, a lower figure compared with that of other studies. Climate change has furthermore been the first foreign policy priority for Spaniards since 2017, especially for respondents on the left and centre of the political spectrum and across all age groups. However, the war in Ukraine has significantly lowered the relevance of climate change while raising energy security to the fourth foreign policy priority.

The abundance of natural resources for transition (sun and wind), a competitive renewables industry and a wide-ranging legislative and executive climate and energy transition framework underpin the resource-research-market-regulation quartet that positions Spain as a potential frontrunner and (currently) a staunch supporter of the EU's net zero goal. The regulatory framework includes: a Strategic Energy and Climate Framework including a recently adopted Climate Change and Energy Transition Law; Spain's (Integrated) National Energy and Climate Plan (NECP); a Long-Term Decarbonisation Strategy; as well as a Just Transition Strategy that includes French-inspired Just Transition Agreements.

The above framework includes ambitious decarbonisation goals, *inter alia*: at least a 23% reduction in emissions of greenhouse gases by 2030, compared with 1990 levels; climate neutrality before 2050; 42% of renewables in final energy use by 2030 (requiring around 6 gigawatts of additional renewable energy production per year); 74% of renewable power by 2030; 39.5% of energy efficiency improve-ment by 2030; and 5 million electric vehicles by 2030.

According to the government, Spain comfortably met its 2020 renewable and energy efficiency targets, the latter thanks in part to lower energy demands during the first COVID-19 lockdown. However, Spain is lagging behind target in the transport sector as well as in building renovations. The RePowerEU boosted EU renewable targets (320 gigawatts of newly installed solar photovoltaic in 2025—over twice the current EU amount—and both 600 gigawatts of photovoltaic and 1.236 gigawatts of renewables in 2030) which could prove challenging for Spain if its renewable targets are enhanced given its slow-permitting environment. There is also rising opposition to large renewable projects in the 'emptied Spain'. These projects are perceived by some as 'green colonialism', not co-designed with the affected communities, potentially affecting the environment and leaving little benefits behind in terms of growth and jobs. As decarbonisation speeds up, it will be increasingly important to address those issues. Building on existing Just Transition Agreements, the government could help design 'Socio-environmental Transition Contracts' to enhance renewable deployment while limiting social backlash.

Reaching the goals enshrined in Spain's NECP is expected to require EUR 241 billion in investments by 2030, 80% of which is expected to come from private sources. The chances of meeting Spain's climate goals and speeding up the NECP have improved with the green-eschewed National Recovery and Resilience Plan that allocates 40% of the EUR 69.5 billion in EU grants to supporting climate goals (where investments with a 'climate tag' of 40% or more are largely allocated to the transport sector—responsible for over a quarter of Spain's emissions—housing and energy sectors).

The overall economic impact of the NECP is expected to be positive, adding 1.8% to GDP in 2030 alongside annual increases in employment (253,000 in 2021, rising to 348,000 in 2030). The largest employment gains are expected in: retail trade; repairs; the building sector; the hospitality industry; scientific and technical activities and transport. Limited transition-related employment losses are likely in the extractive industries, according to the NECP's socioeconomic impact study. As expected, business associations such as the Spanish Green Growth Group that stand to benefit from (or are better placed to deal with) the low-carbon transition support Spain's decarbonisation efforts whereas the Alliance for the Competitiveness of Spanish Industry (including companies in the oil, chemical, cement and paper sectors) has shown greater reluctance towards increasingly ambitious climate goals.

The question remains as to whether or not these goals will be achievable in the aftermath of Russia's invasion of Ukraine and in a context of worsening economic outlook, high debt and inflation. Added to these headwinds are regulatory changes and subsidies to fossil fuels that, while well-intentioned in their goal of sheltering consumers, have had a rebound effect in prices, allowing producers to appropriate part of the subsidies and sending the wrong signal for decarbonisation. Future measures to limit the impact of increasing energy prices would need to be tailored to low-income groups, rather than subsidising the population at large, and be time-limited to avoid distorting the market.

Spain's climate exposure, abundant renewable resources, strong renewable sector, citizens concern about climate change and the current political set-up have all helped upload increasingly ambitious renewable and energy efficiency targets to the EU. On the other hand, EU climate and energy targets are expected to set the least common denominator for future decarbonisation ambitions in Spain across political parties. This was so in drafts of the Climate Change and Energy Transition bills presented in 2018 by the conservative Popular Party, the far-left Unidas Podemos and the socialist party. Except for the far-right group (Vox), it can be argued that Spanish political parties look to the EU as their climate compass. This means cooperating in the implementation of increasingly ambitious decarbonisation targets. Spain nevertheless is known for seeking to nuance Fit for 55 dossiers that are perceived to impact the most vulnerable (such as the expansion of the Emissions Trading System to the transport and building sectors).

Recommendations

Given Spain's current regulatory ambition, citizens concern about climate change, renewable and electric grid management potential, it is argued that the country can further contribute to the EU's energy security and climate leadership in the aftermath of Russia's invasion of Ukraine. To realise Spain's full potential as an EU renewables bedrock, the country will have to end its 'energy island' status by significantly increasing its interconnections. Historical opposition to such interconnections by France would have to be halted for that to happen. Continued support for a Just Transition, speeding renewable permitting processes and establishing procedures for mandatory co-development of renewable projects alongside communities will also be essential levers for Spain's transition capacity potential to materialise.

Lara Lázaro-Touza is Senior Analyst at the Elcano Royal Institute and Lecturer in Economic Theory at Centro de Enseñanza Superior Cardenal Cisneros (attached to Universidad Complutense de Madrid).

The Elcano Royal Institute is a Spanish think tank for international and strategic studies. It is based in Madrid and was created in 2001 as a private foundation. The goal is to foster the creation and exchange of ideas in a plural and independent environment, with a stable and multidisciplinary team of analysts and a wide-ranging network of associated experts. It is also a member of TEPSA.

Sweden: Much Progress—But More Is Needed!

Gunilla Herolf

High Ambitions

The Swedish government takes climate change seriously and aims to reach the goal of net zero carbon emissions by 2045, five years before the EU's deadline. This implies reducing domestic emissions by at least 85% relative to the 1990 levels. Another objective is to have firstly fossil-free and secondly nuclear-free electricity by 2040.

The surplus of emission rights generated has been removed every year since 2014, instead of being sold to other countries. A large majority in parliament shares these goals and an independent interdisciplinary body, the Swedish Climate Policy Council (Klimatpolitiska rådet), continually monitors the policies' adequacy.

The Challenges

Nature has endowed Sweden with the potential for hydropower and the availability of biomass from vast forests. Accordingly, the country has by far the highest share of renewables in the EU (amounting to 60% of energy consumption in 2020). Petroleum products account for 20% and the nuclear share is 10%.

However, geography also creates challenges for a country that is 1500 km long. Hydropower is generated in the north, whereas nuclear power plants, as with most of the population, are mainly in the south. Since nuclear power has been reduced from 12 to 6 plants (but with no political stop date), new sources of renewable energy in the south are needed. Wind energy is central for meeting the goal of fossil-free electricity by 2040 and sea-based wind energy parks are seen as a solution.

G. Herolf (✉)
Swedish Institute of International Affairs, Stockholm, Sweden
e-mail: gunilla@herolf.se

Challenges also lie in some of the traditional industries being large greenhouse gas emitters. Domestic (primarily road) transport and industry (notably metal and cement), each accounting for 31%, together with agriculture at 15% top the list.

According to the Climate Policy Council, Sweden must focus on four key areas: more efficient use of energy and resources; zero emissions electrification; biomass from forestry and agriculture; as well as carbon capture and storage.

Transformation of Key Industries

Even though services account for two-thirds of the Swedish GDP, classic process industry is still crucial for the economy. The Russian war is not relevant for transformation efforts: Oil is bought on the spot market and the Russian imports share of 8% in 2021 can be replaced. Gas (accounting for only 2% of the Swedish energy supply) comes via pipelines mainly from Denmark, half of it from Russia. The Industrial Leap programme was initiated in 2018 supporting innovations to reduce process-related greenhouse gas emissions. Since both technically difficult and expensive for the companies, financial support is provided by the Swedish Energy Agency. Financial assistance also comes from the EU Recovery and Resilience Facility. Sweden will receive EUR 3.3 billion of which 44% is aimed at climate objectives.

Steel producer SSAB accounts for around 10% percent of Swedish greenhouse gas emissions. A pilot plant now operates in northern Sweden, with the goal of having an industrial process in place by 2035, which is totally dependent on renewable energy. Fossil-free steel would be a big step forward, but production also requires a great deal of electricity.

The cement market is dominated by Cementa, which is also a heavy producer of CO_2. Its main strategy is to capture and store CO_2, on which basis the company now claims that it will be climate neutral by 2030. Other climate-neutral techniques are being developed by competitors.

The forestry sector has for decades sought to substitute non-renewable fossil fuel. Wood can replace steel as well as concrete even for high-rise buildings and replace fossil-based plastic as well as textiles. Preem oil refinery, one of the largest emitters, has initiated production of bio-oil, a waste product of forestry and agriculture. The use of biofuels, another forestry residue, has helped to cut down transport emissions.

Applications from renewable energy are now increasingly competitive and northern Sweden has become a hub of industrial processes based on clean energy sources. However, some challenges remain and Sweden must speed up to meet its goals. Road transport and agriculture, for example, need to reduce emissions. Sweden's aim of meeting the enormous future demand for fossil-free electricity (emanating from the needs of new technology and the increasing number of electric vehicles) is deemed to be achievable. The discussion now is rather about whether energy from the expanding wind energy will suffice or whether some new investments in nuclear energy will be needed. Furthermore, the interests of reindeer husbandry and nature protection clash with the requirements of those who want to excavate rare minerals needed for the new technology.

Sweden and the EU

With its close engagement in climate change issues, Sweden sees the EU's role as crucial. It is generally positive not only about the Fit for 55 package, in which the EU commits to cutting emissions by at least 55% by 2030, but also the means of accomplishing this aim.

Sweden has mixed views about the forestry proposals, some positive but others highly critical as far as the means are concerned. A common forestry framework, with increased centralisation and supra-nationality as well as an increased EU budget, are all contrary to the government's thinking. Nor would this be an efficient way forward, since over-arching instructions across the Union would not always suit local conditions. It is further argued in Sweden that by bringing up only certain aspects of forestry and not including the forest's many roles in connection with climate change, these proposals are not suitable for Nordic forestry. Moreover, if implemented, these supra-national rules would make it harder for Sweden to reach its climate goals. In keeping with this issue, it should also be mentioned that the country has been criticised for its policy of 'clear-cutting and monoculture'.

According to Eurobarometer 96.1, 60% of Swedes (the highest among EU Members) see the environment and climate change as the EU's main challenge, as compared with an EU average of 32%. Some organisations and individuals, such as Greta Thunberg, are very critical about the lack of progress and see the government as being too passive. The Swedish Society for Nature Conservation (Naturskyddsföreningen) complains that Swedish reductions are not sufficient and pushes for more radical policies, such as the outright prohibition of fossil fuels. The youth organisation, Aurora, intends to sue the Swedish state over alleged failure to meet its climate change commitments. People are becoming increasingly aware of how urgent it is to take action, which can be witnessed, for instance with rapid increases in the number of electric vehicles and solar heating systems. The present heat wave (also in Sweden) has made it a major discussion theme amongst the entire population.

Recommendations

Understandably Russia's invasion of Ukraine and its potential consequences demand attention at present. However, it must be made clear to all that climate change is interconnected with other threats and that it would be equally disastrous to consign all these pressing issues to the back burner. As a matter of great urgency, countries must spread best practices and seek common solutions with neighbours.

The choices made by millions of people can have extraordinary effects and the importance of popular motivation, therefore, cannot be overrated. Hence, no effort should be spared in demonstrating to people their potential for collective power in bringing about change. As for Sweden, it should use the opportunities presented by its EU presidency (spring of 2023) to push for reforms in order to achieve the EU's climate goals. It is vital that emissions within the European Trading System are scaled down, especially in the case of major emitters. Sweden also needs to tackle the issue of the 'local veto', which has prevented many wind energy parks from being built, even though people in general endorse their use.

Gunilla Herolf is a Senior Associate Research Fellow at the Swedish Institute of International Affairs (SIIA/UI) and a Member of the Royal Swedish Academy of War Sciences (Vice President 2010–14). She was also previously a Senior Researcher at the Stockholm International Peace Research Institute (SIPRI).

The SIIA/UI is an independent institute for research, analysis and information, founded in 1938 as well as being a member of TEPSA.

Ambitious Dreams Versus Harsh Reality: Can the Netherlands Really Become a Frontrunner in Climate Action?

Niklas Mayer

The Importance of Climate Change Action in Dutch Politics

Being one of the lowest-lying and at the same time one of the most densely populated countries in the world, the Netherlands is by its very location extremely vulnerable to the effects of global warming and consequent rise in sea level. Experts believe that the country would manage to compensate for a rise of up to two metres with its engineering expertise together with the use of dikes, sea walls and other constructions. However, anything more would result in relocating complete cities and giving up whole regions. To say the least, this is exceptionally worrying given some predictions that the Netherlands could face a sea level rise of three metres by the end of this century.

The new Dutch government was sworn in during January 2022, after negotiations lasting an unprecedented 10 months. COVID-19 and climate change action are top priorities for the cabinet Rutte IV, with the latter being one of the decisive topics over the course of the election campaign.

Well aware of the climate crisis, Dutch society frequently pushes the government to scale up its actions. Indeed, the formulation of ambitious emission reduction goals of 49% by 2030 came about as the direct result of civil engagement. The Urgenda Case in 2015 was the first case globally of citizens opposing their government on the matter of climate change. The Dutch Supreme Court sided with the citizens movement—represented by the environmental foundation Urgenda—and ruled that the Dutch government should increase their climate change action so as to bring about a 25% reduction in national greenhouse gas emissions by 2020. This so-called Urgenda target was subsequently accomplished, albeit with additional assistance from the COVID-19 pandemic. This ruling had further been extended

N. Mayer (✉)
Maastricht University, Maastricht, Netherlands
e-mail: n.mayer@maastrichtuniversity.nl

115

by the appeal court in 2018 and the supreme court in 2019. It was those rulings, triggered by ongoing civil activism and the high priority given to climate change issues by Dutch society, that pushed the Rutte III government to prepare and pass the Dutch Climate Agreement.

Internationally, the Netherlands increasingly wants to take a leading position in climate change action, despite being one of Europe's largest polluters. In this context, the last cabinet of Prime Minister Mark Rutte upheld complete implementation of the Paris Agreement, pushed for EU-wide environmental targets as well as the Green Deal and presented the Dutch Climate Agreement in 2019.

Key Industries, Climate Advocates and Opposing Forces

Some of the Netherlands's key industries are, as just implied, highly polluting. This includes agriculture, food processing, steel, chemicals, as well as oil and gas exploitation along with refineries. In this context, an important example of the battle between forces opposing climate change action in the Netherlands and climate change champions (motivated advocates) is the case of Royal Dutch Shell vs. Roger Cox. The Dutch gas and oil company (now also British) is responsible for 2% of all emissions globally. Roger Cox, lawyer for Friends of the Earth Netherlands, achieved a milestone in history: for the first time a private company was required to comply with the Paris Agreement. More specifically, Royal Dutch Shell is mandated to achieve a 45% reduction in its emissions by 2030.

Another example of vibrant civil support for ambitious climate action is Aniek Moonen, Chair of the Young Climate Movement (Jonge Klimaatbeweging), whose organisation aims to bring the younger generation's voice into Dutch policy-making. In this undertaking, the movement is certainly gaining weight and frequently in discussion with ministers and high-level politicians.

At the same time, there is large potential here for on-shore and off-shore wind energy production. If this is further expanded with decarbonisation of electricity and energy consumption, much of the domestic and industrial greenhouse gas emissions can be cut. The majority of urban areas, as well as major industries—including refineries, chemical and steel production plants—are close to the coast and could be supplied directly by green off-shore energy.

As to Dutch energy imports, in the last 10 years Dutch domestic energy (gas) production has decreased significantly, resulting in an increased dependency on imported energy sources. Indeed, since 2018 the Netherlands has been a net gas importer. As to the energy mix, this comprises natural gas (38%), oil (35%), coal (10%), nuclear, solar, hydropower and wind together (11%), along with biofuels and waste (6%).

Almost all of the coal is imported, coming from Russia, the USA and Colombia. The main countries of origin for Dutch petroleum gas imports are the US, Belgium and Norway. The trading partners for crude oil imports are Russia, the UK and the USA. Whilst there is currently some energy dependency on Russia, Prime Minister Rutte estimates that this can be completely phased by the end of 2022.

The Netherlands as a Frontrunner in EU Climate Action?

At EU-level, the Netherlands is famous for being a member of the 'frugal four' group, together with Austria, Sweden and Denmark (along with Finland which frequently joins this group). These states are firmly against an extended EU budget and common EU-wide debts. Furthermore, during EU budget negotiations for 2021–2027, the frugal four argued that at least 25% of the budget should be spent on climate change action.

This fits in well with a picture of the Netherlands generally being more committed in this regard than most other EU Member States, by trying to make the EU's common environmental and decarbonisation goals more ambitious.

In this context, the last Dutch government (Rutte III) was among those ambitious EU Member States who lobbied for—and finally achieved—an adjustment to the 2030 EU emission reduction goals from minus 40% to minus 55% compared with 1990 levels. One of the Netherlands key financial tools for implementing both the Paris Agreement and the Dutch Climate Agreement is the SDE++ scheme, which finances renewable energy as well as decarbonisation technologies. Since subsidies under the SDE+ and SDE++ schemes show impressive success when it comes to convincing and supporting industry to undertake decarbonisation processes, these schemes might become a best-practice example for other EU states.

The Dutch Climate Agreement, just mentioned above, is aligned to the EU's Green Deal and Energy Governance mechanism. At national level, this Agreement also calls for 10-year plans, with stocktaking and revision mechanisms to be applied every five years. These plans will outline ways of reducing emissions, by decarbonising energy consumption and electricity through the upscaling of renewable energy and other energy sources, as well as being generally more energy efficient. This approach of 10-year plans with control and revision mechanisms is identical to the EU's National Energy and Climate Plans, as introduced by the Clean Energy for all Europeans Package 2018.

Since some goals in the Dutch National Climate Agreement are more ambitious than EU policy, the Dutch heavy industry has real fears about losing its competitiveness, especially with a view to the Dutch National Carbon Tax. This tax is part of the 2019 Dutch Climate Agreement and is one of the few worldwide to tax emissions from heavy industry. Purely from a climate perspective, this is a promising step towards Dutch climate neutrality by 2050. In this regard, the Dutch government expects that the levels of scope and ambition implied within the EU Emissions Trading System will be upscaled significantly, so that the Dutch National Carbon Tax for heavy industry would either become unnecessary, or at least reduce the competitive disadvantage that Dutch companies would face, when compared with industry elsewhere in the EU.

Path Towards Climate Neutrality and Recommendations

Thanks to a favourable public perception towards Dutch climate change action,[1] as well as the court rulings against the Dutch state and the private company Royal Dutch Shell, important policy changes have been implemented and the new government of Mark Rutte IV seems determined to be Europe's frontrunner on climate change action, whilst at the same time maintaining its industrial competitiveness.

Since the Netherlands is currently still one of the EU's major polluters (over 10 tonnes of CO2 per capita compared with the EU average of 7.5 tonnes) and much of the Dutch economy relies on polluting industry, it remains to be seen if the country achieves this turn-around and the ambitious 2030 goals: 49% emission reduction compared to 1990 (or even 55% according to the upscaled EU goal). In other words, the Netherlands is currently a frontrunner in theory (in agreed policy goals). In practice, though, with regard to emissions the country is still far away from being a role model.

These developments in the Netherlands, if achieved, could have positive spillover effects on other EU countries. If the country delivers on its ambitious plans and overcomes the challenges, it can be seen as a best-practice example, thereby motivating other EU states to do the same. In this context, it is of paramount importance that the EU supports Dutch goals and especially the decarbonisation of Dutch heavy industry. Next to extension of the EU Emissions Trading System mentioned earlier, Dutch industry should be supported with EU funds to help it meet the challenges of green transition. This has to include the expansion of renewable energy. Because of elections and very long coalition-building negotiations, the Netherlands was the only country by January 2022 that had yet to submit its recovery programme to the Next Generation EU recovery fund. The EU should use these funds to finance green transitions in the Dutch industries mentioned earlier.

As to the government of Prime Minister Mark Rutte IV, its priorities should revolve around rigorous implementation of ambitious plans outlined by the Dutch Climate Agreement. Many of the points from this agreement have still to be transposed into law. Moreover, green transition will be very demanding for all major Dutch sectors, specifically: steel and chemical industries, as well as agriculture. Furthermore, energy efficiency in offices and domestic housing, as well as the expansion of renewable energy sources will create enormous challenges. While national subsidies for companies who implement reforms to decarbonise have already shown some beneficial effects in the steel and chemical sectors, such subsidies should now be expanded to agriculture and the energy efficiency of buildings.

---

[1]Even though some parts of Dutch society, such as farmers, increasingly demonstrate against stricter climate policy.

Niklas Mayer   is a research assistant to Dr. Giselle Bosse's Jean Monnet Chair in EU Politics in a Changing Global Context ('Chance'). Niklas' PhD research revolves around the impact of climate resilience-building projects on migration decisions in Ethiopia. At Maastricht University (Netherlands), Niklas designed and teaches the course 'EU International Relations and Climate Change'. Prior to that, he obtained a master's degree in European Studies from the College of Europe and a Bachelor's degree in International Relations from the Autonomous University of Madrid.

Maastricht University is the most international university in the Netherlands and distinguishes itself with its innovative education model, international character and multidisciplinary approach to research and education.

Part II

EU Neighbours

Albania's Challenges and Risks on Climate Neutrality!

Edmond Hoxha

Albania is a small country with an economy led by the service sector, alongside some industry and agriculture. Many challenges result from the impact of climate change, in particular extreme weather conditions which are creating not only heavy flooding, landslides and tidal waves, but also drought. Added to this, COVID-19 and Russian aggression against Ukraine is profoundly affecting Europe. The impact of these extreme events has also reached Albania, with increases in fuel prices and consequently other products immediately becoming unaffordable for the vast majority of citizens. Fortunately, Albania uses mostly water energy as well as some solar power and hence there is no heavy dependence on Russian gas, albeit the general energy crisis has caused a national emergency.

Combatting climate change is one of the Albanian government's key priorities, but regrettably efforts towards environmental protection are in practice behind the rhythm of economic development. However, it is important to stress that whilst 90% of the country's energy needs are supplied by hydropower, which is clearly of great benefit, serious problems also exist in regard to water resources and agriculture irrigation. Considering this situation, the government is promoting other forms of renewable energy, such as solar and biomass.

Albania has joined European endeavours to be the first climate-neutral continent in the world by 2050. Currently, the government has adopted Legal Acts related to climate change, with the National Energy and Climate Plan aimed at reducing greenhouse gas emissions, adapting to climate change, as well as unifying Albania with other countries that recognise the climate emergency and contribute to global efforts in the fight against climate change. Civil society is also active in protecting the environment, but whilst its recommendations concern important issues, they largely receive little consideration from the Government.

E. Hoxha (✉)
Polytechnic University of Tirana, Tirana, Albania

Combatting climate change is undoubtedly crucial for Albania. Whilst it is true that using coal-fired thermal power plants for electricity generation has long been abandoned, transition to climate neutrality will require large investments, special resources and dedicated projects. Given this situation, Albania should see decarbonisation, transition towards 'Green Energy' not only as a challenge, but also as the provider of employment opportunities and new business models related to renewable energy.

Albania has already successfully used water energy and should now focus on solar and wind energy. Some solar plants are already active, but extending the use of this source needs high initial investment, which most people find simply unafford-able. Although some financial support schemes have been developed, they are insufficient and thus the use of solar energy is still low.

Transition to a climate-neutral economy requires Albanian politicians to address certain key issues such as:

- Providing the necessary support to ensure that economic growth will not take place at the expense of ecosystems and natural resources which are in decline.
- Raising public awareness on climate neutrality.
- Planning support for the transport, industry, agriculture and forestry sectors that will be affected by climate change.
- Ensuring the conservation of biodiversity at sea and in rivers.
- Supporting poor and needy families that are most likely to be exposed to drought, flood and extreme heat.
- Creating new jobs, together with redeploying the workforce and their retraining.

Balancing short and long-term benefits for companies, which are reluctant to invest in climate neutrality due to cost and unfair competition, the government must find ways to motivate them through low taxation, grant support and so on. Similarly, it must allocate investments in areas with high transition decarbonisation costs, whilst at the same time achieving economic and climate objectives. In regard to regional characteristics, the model chosen should be tailored accordingly by under-taking regional studies in order to adapt to the so-called 'Fair Transition Mecha-nism'. It should address support for Global Climate Action as a priority for Albania's Foreign Relations, namely 'Climate Diplomacy' in accordance with the Paris Agree-ment. This means much closer cooperation at international, regional and domestic levels.

Transition to a climate-neutral economy will have a major impact on Albania, both from financial and social perspectives. It is already known that the main global contributors to climate change are: transport; industry; agriculture; and fossil-based energy production. It is difficult to give exact percentages of greenhouse gas emissions for each sector, but transport is certainly the main contributor, followed by construction. Accordingly, the country's most affected sectors will be transport, construction, agriculture and mining. In the transport sector, existing petrol- and diesel-powered vehicles will need to be replaced by their electric equivalents. This transformation is likely to meet fierce resistance from businesses as well as the general public, particularly poor and middle-income groups.

Whilst some solutions are already available, the general public and businesses need to be aware that transition to a climate-neutral economy will be difficult, but an absolute and vital necessity in safeguarding the well-being of people. Financial support or compensation for business losses due to technology changes must be associated with retraining employees and providing long-term loans for the purchase of new technologies that rely on renewable energy. Moreover, households must receive financial support as a means of stimulating the use of solar panels, with any differences in price increases due to changing technologies being compensated for and traditional equipment which has been made redundant being subject to free collection by the state.

A survey on lost and recoverable jobs should be undertaken, with redistribution according to needs and regions. Free retraining courses must be made available for employees who leave without a job together with financial assistance for up to three years. Construction companies must be mandated by law to rely on renewable energy for each new project, providing the necessary amount of energy needed for heating and hot water from the solar panels installed in each building.

The general public is less aware and consequently less active in climate change issues. Most ordinary people do not understand the consequences of climate change and furthermore, they find it difficult to understand the concept of 'climate neutrality'. Albanians generally are far more concerned with issues such as the weak economy, low salaries, high prices and taxes, corruption, property problems and justice. To a large extent, climate neutrality is understood only by environmental experts and academics, but even their responses are lacking. The most active in this regard is that part of civil society which deals with environmental issues. Currently public perception of the climate crisis is more related to an energy crisis with its associated general increases in prices for energy, fuels and other products.

Another important part of public perception is related to the EU's support for Albania. On the one hand, people believe that the EU perceives Albania as only a very small contributor to greenhouse gas emissions globally. On the other hand, the EU also knows that Albania is very vulnerable to climate effects and hence there are steps which are necessary to be taken in achieving 'net zero' greenhouse gas emissions. Despite this, so far there has been little EU support for any climate change impact.

In general terms, Albania's relationship with the EU is not driven by environmental issues or climate change, simply because the country is not a major contributor to greenhouse gas emissions. Of course, climate change is an issue that affects all European countries, whether or not they are members of the European Union. In this sense, all countries must work together. For Albania, cooperation with the European Union is at a very good level, albeit the integration process is extremely slow. The country is experiencing a very long transition process which has so far lasted 31 years. This is both incomprehensible and disappointing for the Albanian public, so much so that even the Albanian people's great desire to become part of the EU, which was about 90% in the 1990s, has now dropped considerably. It seems that EU integration has become a mobile target. Every time Albania fulfils its obligations, at the last moment a new target is set which once again delays its ultimate accession. Whilst the country's support for climate neutrality objectives will undoubtedly strengthen relations between Albania and the EU, this alone is not expected to result in rapid membership. However, an important step in the right direction was taken in 2020 when formal EU accession negotiations were opened with Albania.

Albania has so far supported every EU initiative, even those that exceed the abilities of a small country with a weak economy and limited resources. The Albanian Government has many ambitions in this process, but in my opinion, there are three principal issues. Firstly, the Albanian Government expects acceleration of its EU membership process, especially after Russia's invasion of Ukraine. Secondly, this whole delicate and in some ways painful process for the citizens needs strong external support. Albania is not a contributor to carbon pollution and thus deserves assistance through EU Public Climate Financing, primarily to curb rising prices and finance the use of solar energy. Thirdly, transition to a 'new system' will require qualified expertise and retraining for members of the workforce who will be left jobless. This means that in the long run support from the Education System and Scientific Research will be needed for new concepts, methods and technologies oriented by climate neutrality.

Recommendations

In light of what has been covered above, I conclude that Albania is seriously affected by climate change. Whilst its inclusion in the major EU objective of making Europe the first climate-neutral continent represents an enormous challenge, it is certainly the right direction to safeguard the continent's population.

In order to advance this process, I would recommend the following measures: (1) raise public awareness of the difficulties and benefits of this objective; (2) set out a National Energy Plan with concrete measures and deadlines, prior to which a Regional Cooperation Plan should be drafted to achieve this objective; (3) conduct a national survey on the 'Impact of Decarbonisation' (jobs lost, jobs created and redistribution according to needs and regions), involving the University, civil society and social partners; (4) provide financial support or compensation for business losses due to changing technology; (5) support families with grants for solar panels and reimburse price differences due to changing technology; (6) offer free retraining courses for employees who lose their jobs and provide financial assistance for up to three years; (7) support the education system and scientific research in the development of new concepts, methods and technologies oriented by climate neutrality.

Edmond Hoxha received his PhD degree on Geosciences and Environment at the Polytechnic University of Tirana, Faculty of Geology and Mining. He also studied 'Leaders on Development' at Harvard University, USA. He has considerable experience working with the Government of Albania and International Institutions such as the World Bank, European Union and GIZ. During 2009–2013, he was Deputy Minister for European Integration of Albania. He is founder of the Albanian Centre of Excellence, the National Mining Surveyors and Geomatics Association and publisher of the scientific journal Albanian Excellence. He is founder of the Western Balkan Conference on GIS, Geodesy and Mine Surveying. Currently he is Professor in the Faculty of Geology and Mining teaching GIS Technology, Mine Modelling and Project Management.

The Polytechnic University of Tirana is the oldest and the second largest university in Albania, after the University of Tirana. It was founded in 1951 and now has approximately 10,000 students, who come from Albania, Kosovo, Montenegro and North Macedonia.

Mission (Im)possible—How to Fight Against Climate Change in a Country Enduring Permanent Crises: The Case of Bosnia and Herzegovina

Vedran Dzihic

Citizens of Bosnia and Herzegovina quite often take their demands to the streets. It might be unusual, but nevertheless understandable given the mounting political, social and economic problems that the country has been facing ever since the end of the war in 1995. Some of the most prominent recent protests to emerge have been concerned with environmental issues, primarily the destruction of Bosnian rivers or horrific air pollution in Bosnian cities. One of the most prominent protests in recent years was titled 'The Brave Women of Kruščica'. During 2017 a number of women from the small Bosnian village of Kruščica organised themselves and went on duty for more than 500 days, blocking the local bridge, preventing the passage of excavators and thus the construction of mini hydropower plants on the river which would produce negative environmental impacts. Despite the authorities having decided to send in police forces that were accustomed to using brutal physical force against activists, these courageous women eventually succeeded with the Cantonal Court in the city of Novi Travnik annulling construction and urbanist permits, thereby halting the construction of two planned mini hydropower plants. This fight inspired a group of young activists in the city of Jajce to organise large-scale protests against investors in small hydropower plants on the beautiful Pliva River. The plan by investors and local political authorities to build two small hydropower plants only a few hundred metres from the world-famous Pliva Waterfall in Jajce was seen as a direct attack on the local community and natural resources.

Following a range of similar initiatives, an advocacy group of activists started a campaign aimed at changing the law so as to ban the construction of small hydropower plants. After ten years of pressure and struggles by activists in numerous local communities and more than 60,000 signatures collected during the public campaign, the activists finally prevailed. At the beginning of July 2022, the upper house in the

V. Dzihic (✉)
Austrian Institute for International Affairs, Wien, Austria
e-mail: vedran.dzihic@univie.ac.at

Federation of Bosnia and Herzegovina entities parliament confirmed changes to the entities' Law on Electricity to ban the building of small hydropower plants because of their negative impact on the environment.

The fight against small hydropower plants has to be seen in the broader context of climate change, which is hitting Bosnia and Herzegovina similarly to all other European countries. Fighting for natural resources as the backbone of tackling climate change in general is one of the ways to raise citizens' awareness, which to date is still very limited. The country desperately needs these champions who are prepared to fight environmental destruction and thus contribute to the fight against climate change.

While citizens and local communities suffer under climate change and fear the destruction of natural resources with many deciding to protest, the authorities usually remain silent and reactive. Alternatively, they may try to address the issue only through very formalised national action plans and strategies with no concrete implementation thereafter by way of follow-up. In general, climate change is not perceived as a high-priority issue. Amidst constant political and social crises that have been haunting the country for decades, attention to the rather slowly unfolding climate change is rather limited.

According to latest opinion polls conducted by the International Republican Institute in the spring 2022, the prominent problems here are related to the economy—unemployment and rising prices—and corruption. Among the top ten most important problems for citizens, the environment and climate change do not even feature. Looking at these and similar opinion polls, one must conclude that average people in Bosnia and Herzegovina simply do not have enough time and space to be worried by the impact of climate change above all other types of crises that dominate their lives, such as corruption, unemployment and poverty, which are much more visible and palpable.

Yet, the real impact of climate change on Bosnia and Herzegovina is huge. According to the World Bank's Climate Risk profile of the country, it is certain that projected climate changes will make the country increasingly vulnerable to natural hazards, including heat waves, droughts, floods and landslides. Bosnia has been experiencing a steady rise in temperatures with the most pronounced increases occurring over the past 30 years. There is also a high risk of forest fires due to projected temperature increases; indeed, during a period of extreme heat during the summer of 2022 a number of forest fires broke out. At the same time, extremely heavy rainfalls have become more frequent, which has led to major seasonal flooding. The country was, for example severely hit in 2014 with floods unprecedented in 120 years of weather records. This affected around 25% of the population, severely disrupted the economy and inundated 81 municipalities as well as agricultural fields. It triggered more than 3000 landslides, bringing about a fall in GDP of almost 15%.

Added to these direct effects of climate change comes the question of the country's energy supply, which presents a delicate problem for the planning of any efficient strategy to fight climate change. Bosnia is highly dependent on coal and its energy production is responsible for over 60% of greenhouse gas emissions in the country, thereby largely contributing to one of the worlds' most pressing air pollution problems. With thermo-power plants being responsible for more than 65% of electricity in Bosnia, it is quite obvious that any energy transition has to start with reducing this dependency on coal and moving towards sustainable sources of energy. Following Russia's invasion of Ukraine and subsequent energy crises throughout Europe, it is rather unrealistic to expect that Bosnia will soon be able to make any progress towards sustainable sources. On the contrary, at least in the short-term plans for additional coal power plant investments (particularly in the project Tuzla 7 financed by China) might again become relevant.

The country was supposed to start its energy transition process a long time ago not only to draw closer to the Paris Agreement goals and the regionally relevant Western Balkans' Green Agenda (based on the Sofia Declaration signed in 2020), but also to reduce its own emissions and contribute to mitigating the climate crisis. The goal set is for decarbonisation by 2050. In this process authorities can count on the EU's support. The recently launched Platform Initiative in Support for Coal Regions in Transition in Western Balkans and Ukraine promises to mobilise at least EUR 100 billion to support the shift from coal to a greener economy. One tranche, some EUR 9 billion, will be available to the countries of the Western Balkans. The new Instrument for Pre-Accession Assistance (IPA III) with a total budget of over EUR 14 billion for the period 2021–2027 is largely related to the Green Agenda.

Incentives and support are certainly available now, but so far in regard to implementation of the very ambitious plans not much has happened. At a formal level the authorities on all levels of governance (state, entities, cantons) do invest considerable effort in formulation of policies and adaptation of national action plans. But as is the case with almost all other reforms in Bosnia related to the EU integration process, climate change reform efforts are in writing only with no serious follow-up and implementation plans. It is precisely because of these serious setbacks for the process of EU integration in general as well as the general lack of interest for climate change-related topics that the Bosnian population does not place any partic-ular hopes in the EU as an agent of climate change for the country.

Recommendations

The obvious political stalemate in Bosnia and Herzegovina comprises self-consumed political elites, those quoting ethnopolitical argumentations and political representatives, all of who lack any focus on topics of relevance for the population. Accordingly, this poses the biggest problem for the country's functionality in general, as well as the changes and adaptation needed specifically to meet the challenges of climate change. Indeed, a switch in political power would be needed to free the potentials for EU-related reforms and increased functionality in the country. Only through a new political set-up and the necessary investment could the country start to reverse the majority of citizens lack of awareness about the importance of green energy for the future of society. Such a new context could bring about a stronger focus on implementation of existing laws and commitments that extend beyond mere promises and result in real change. Obviously, all these efforts need to merge into the wider regional framework and efforts to address climate change-related issues together with neighbours and other countries in the Western Balkans with support from the EU and other relevant international organisations.

Finally, fighting climate change as well as environmental destruction and working towards green transition is a topic that goes beyond political fights and ideological labelling. This must become Bosnian society's key focus. Activists mentioned at the beginning of this chapter do engage in actions precisely on these grounds, in other words focussing on problems that need to be solved, leaving behind any ideological or ethnic labels and boundaries, which regrettably are plentiful in Bosnia. It was during the floods in 2014 that Bosnians of all ethnic groups went to help their neighbours from other ethnic communities. They showed a solidarity that transcended any ethnopolitical conflicts of the past. Ultimately, it will be the daily fight of those Bosnian citizens that stepped up in the floods 2014, went on to protect their rivers in Kruščica and in Jajce or mobilised themselves for a general ban of mini hydro power plants who will determine the future of their country.

Vedran Dzihic is a Senior Researcher at the Austrian Institute for International Affairs (OIIP), Co-Director of the Centre for Advanced Studies, Southeast Europe and Senior Lecturer at the Institute for Political Sciences, University of Vienna. He is a non-resident Fellow at the Centre for Transatlantic Relations (SAIS), John Hopkins University, Washington D.C. Dzihic is the author of 4 monographs and editor/co-editor of 15 edited volumes/books. He is also the author of numerous book chapters and scholarly articles.

OIIP is a Think Tank founded in 1979 and committed to fundamental research in the field of international politics. It is Austria's leading institute on international politics at the juncture between academic and policy-oriented research.

Georgia: It Is Time to Address the Devastating Effects of Climate Change

Tinatin Akhvlediani

Impact of Climate Change

Climate change effects in Georgia are readily apparent with increased cases of extreme weather, such as landslides, floods, rising sea levels, storms and other ecological disasters. According to the World Bank, such conditions can be associated with 6% of national GDP loss annually. Half of such loss stems from air pollution's threat to mortality, which in 2018 amounted to the deaths of 4000 people, one of the highest rates in Europe. Air pollution results mainly from transportation and dust linked with the booming construction sector in Georgia. Moreover, the population is exposed to lead poisoning, which results in decreased cognitive abilities, particularly among children.

The effects of climate change are further aggravated by unsustainable use of natural resources, that not only generates direct economic costs, but also decreases future production capabilities. According to the World Bank's 2018 estimates, land degradation gave rise to an 8% loss in agricultural GDP, with a further 5% loss being suffered because of coastal degradation. While the country has unique landscapes, forest, coastal and marine resources, Georgia's legislation covering nature conservation is outdated and hence adoption of a new Law on Biodiversity is vital to avoid any further damage to protected areas.

T. Akhvlediani (✉)
PMC Research Center, Tbilisi, Georgia
e-mail: tinatin.akhvlediani@ceps.eu

© The Author(s), under exclusive license to Springer Nature Switzerland AG 2023
M. Kaeding et al. (eds.), *Climate Change and the Future of Europe*, The Future of
Europe, https://doi.org/10.1007/978-3-031-23328-9_30

Georgia's Climate Policy

Since 1994, Georgia has been a Non-Annex I Party to the United Nations Framework Convention on Climate Change, subsequently joining the Kyoto Protocol in 1999 and hence becoming party to the Paris Agreement in 2007. The country updated its commitment on nationally determined contributions in 2020 to achieve an unconditional reduction in its total domestic greenhouse gas emissions by 2030 to 35% below the 1990 levels and from 50% to 57% below the 1990 levels conditional on receipt of international support. Along with the updated nationally determined contributions, the country has also put forward a 2030 Climate Change Strategy, which sets targets for limiting emissions by 2030. Based on 2015 levels, the Strategy foresees particularly emission reductions in transport, energy and transmissions sectors by 15%, industry by 5% and forestry by 10%. These sectors match those selected in 2015, when greenhouse gas emission levels were recorded as transport 24%, energy and transmission 21%, industry 18% and forestry 32%.

However, this strategy focuses solely on mitigation and does not propose putting forward any national funds to address climate change. Hence, reliance is placed on international support, existing grant and loan programmes together with private sector investment, which has yet to be mobilised. The National Energy and Climate Plan as well as a Long-Term Low Emission Strategy are still being developed.

Cooperation with the EU

Georgia is committed to approximating EU legislation on environmental and climate policies under the Association Agreement. Yet, as the Agreement was concluded with the EU some years ago in 2014, some of Georgia's commitments are now outdated as the EU has moved ahead with upgrading its legislation in line with more ambitious climate policy. In addition, the country has been part of the Energy Community since 2017, which further obliges Georgia to undertake reforms in the energy sector. This refers *inter alia* to the need for significant increases in energy efficiency and the use of renewable energy sources.

According to the European Commission's latest Annual Implementation Report on Georgia, action in the renewables energy sector is still at an early stage, not exceeding 39%, while progress on energy efficiency and climate are moderately advanced, surpassing 51% and 58% of commitments foreseen under the Commission's targets. The country mostly lags behind in adopting a number of Acts for promoting the production and use of renewable energy, improving energy efficiency and promoting a circular economy. The laws on Energy Efficiency in Buildings and Labelling Regulation are already in place, but their implementation is lagging behind as the country has yet to implement a large number of by-laws needed for their enforcement. Besides the legislative basis, implementation of these new laws in practice is regarded as a costly process, which could be burdensome for the private sector. Yet, the national funds needed to ease transition to more energy-efficient practices are not yet available.

Energy Sector

The energy sector in Georgia is responsible for the highest emissions, followed by agriculture, production and industry, along with waste. This is linked to the country's dependence on imported gas and oil, mostly from Azerbaijan, that make up around three-quarters of its energy mix (45.9% gas and 27.6% oil in 2020). The country is rich in hydro resources, but hydropower as yet covers less than a quarter of the country's overall energy demand. In terms of other renewables, such as solar and wind energy, the country's potential is still not exploited due to infrastructure deficiencies. Russia's full-scale invasion of Ukraine has still not mobilised the Georgian government to take further steps towards increasing the use of renewables and phasing out Russian energy supplies. This is extremely concerning, given that the country has registered drastic increases in Russian gas imports since the COVID-19 pandemic's outbreak. In 2016–2020 imported Russian gas amounted to 10% of total gas imports, while by 2021, this share had increased alarmingly to 23%. Since Russia's full-scale invasion of Ukraine, further increase is registered in imports of petroleum products from Russia, which grew by around five times during March-September 2022.

Public Awareness

Public awareness and understanding of Georgia's climate policy as well as the benefits of transitioning to a greener and climate resilient economy is very low. According to a survey conducted within the EU4Climate Programme in 2020, almost all Georgian respondents (97.6%) have heard about climate change and its negative effects. However, only a very small share of the public pursues environmentally-friendly actions and as many as 36.19% believe that they have no chance of making any difference to climate change. In addition, public awareness of the country's climate change policy is rather limited and only around 8% of respondents are aware of Georgia's own policies and commitments. Whatever awareness does exist is higher among men than women and among young people than people aged 65 or older. Other demographic indicators are not significantly linked to climate change awareness.

Fighting climate change, though, is neither on political parties' agendas nor those of civil activists. The Georgian public, political elites and civil society are mainly focused on concerns related to Georgia's European future, receiving a candidate status and high poverty unemployment rates in the country, without linking these concerns to the country's sustainable European integration in line with the EU's Green Deal.

European Green Deal

The EU's objective to make Europe climate-neutral requires close cooperation with its neighbours, particularly members of the Energy Community. Together with its Eastern Neighbours, including Georgia, for 2018–2022 the EU has in place its programme EU4Climate to help Eastern Partnership countries to improve their climate policies and take actions against climate change. In line with its Green Deal proposals, green transformation is among the top priority areas in the EU's Eastern Partnership policy beyond 2020. However, as yet the EU has not specified exactly how it is willing to integrate its Eastern Partnership's neighbours into the Green Deal.

Recommendations

There are a number of actions which Georgia must initiate in order to make progress on development of positive plans towards its achievement of energy goals and green transition.

Climate change effects should not merely be mitigated, but rather more ambitious plans should be put forward and combine with strategic documents to address key issues, including the National Energy and Climate Plan and a Long-Term Low Emission Strategy. Improvement is also needed in managing the use of natural resources, including landscape, forests and coastal zones, by investing in prevention and monitoring systems as well as tools to forecast and address ecological disasters.

Waste management must be updated and the necessary infrastructure must be put in place for more sustainable use of natural resources. Pollution management must be developed, particularly in targeting emissions from the energy sector, transportation and dust from the booming construction sector. Exposure to lead contamination has to be decreased.

Higher investment must be committed to building much needed infrastructure for accelerated use of renewable energy. The country should tap into the potential of Black Sea offshore renewables, possibly alongside hydrogen and electricity grids, while pursuing its decarbonisation objectives and phasing out gas imports from Russia.

Adopting all necessary by-laws must be accelerated to ensure implementation of energy efficiency and labelling laws in line with its Association Agreement and Energy Community commitments. National funds should be made available for promoting implementation of the Climate Change Strategy and the EU acquis in practice across all sectors.

Promoting a circular economy and creating green jobs at the local levels must be undertaken, including those associated with nature and coastal-based tourism. This will also contribute to counteracting reduced growth and the poverty gap across rural and urban areas.

Increasing public awareness about the financial and health benefits of transitioning to green and climate resilient economic activities is vital so as to receive more public support in implementing much needed climate policies in the country.

Finally, the EU should be approached with a view to prolonging its EU4Climate beyond 2022 and most importantly, it should put forward a clear strategy on how to incorporate its Eastern neighbours, including Georgia, in its Green Deal. While some of the Eastern Partnership countries may show hesitation, the EU should focus on helping those of its neighbours who show their political will in seeking transition to low-carbon economies. This would ensure simultaneous transition throughout the region and would avoid carbon leakage which results from placing domestic issues above carbon electricity production and carbon-intensive imports.

Tinatin Akhvlediani is a Research Fellow and the Head of Financial Markets and Institutions Unit at the Centre for European Policy Studies (CEPS) and an affiliated Research Fellow at PMC Research (PMCG). Her expertise and publications cover the EU's financial, trade and neighbourhood policies, with a specific focus on the Eastern Partnership. Her research includes academic papers, book chapters as well as policy studies and evaluations which have not only been delivered for the European Parliament, European Commission and a number of international organisations, but also presented at international conferences in Europe as well as Canada, the USA, China, Australia and Thailand. Tinatin is a frequent commentator at major media outlets, including the BBC, Euronews, Bloomberg, Deutsche Welle, Voice of America and all the main TV channels in Georgia. In 2020, she was listed among 30 Under 30 in the category of science by Forbes Georgia.

PMC Research was founded in 2010 by senior advisers, former government executives and experienced civil servants based on PMCG's experience. It undertakes studies in the fields of economics, politics, energy, good governance and social security. By combining global and local expertise, the Centre elaborates on research-based policy options focused on economic development, accountable and transparent democratic governance. The Centre is also a member of the Trans European Policy Studies Association (TEPSA).

Icelandic Climate Politics: Ways Forward to a Green and Socially Inclusive Welfare State?

Gunnhildur Lily Magnusdottir

Various overlapping crises, climate change, conflicts around the world along with the COVID-19 pandemic seem to be symptomatic of this current era. All serve to highlight the commonalities of crises and how social differences such as age, class, gender, location and education mean that not all people are affected in the same way. If these social differences are not properly addressed in the context of European climate politics, socioeconomic inequalities might be intensified to the extent that ineffective approaches could give rise to protests among groups who feel unjustifiably challenged by climate policies being adopted in Europe.

Historically, the debate on climate change and social differences has focused on vulnerable groups, primarily women, in the Global South. This debate has highlighted the slow violence of climate change, thus the often-devastating effects of climate change on different social groups. In recent years, awareness about the meaning of climate justice and climate-relevant social differences, such as age, gender, location and class, in a European context, including Iceland, has however increased. This is seen, for instance, both in a more people-centred approach in the European Green Deal and in the latest assessment report of the Intergovernmental Panel on Climate Change, stressing the intersectional nature of climate change.

At first glance climate-relevant social differences might appear more to do with many other European countries rather than Iceland, a country which is often portrayed as a relatively homogenous society with small social differences. This is to some extent correct in a European context. However, as the discussion below demonstrates the urban-rural dimension is extremely important since Iceland is a small nation in a country with a relatively large land area.

G. L. Magnusdottir (✉)
Department of Global political studies, Malmö University, Malmö, Sweden
e-mail: gunnhildur.lily.magnusdottir@mau.se

139

Opportunities: Renewable Energy, Well-Being Economy and Social Inclusion

Recent surveys exploring behaviour and views on climate change indicate that over 80% of Icelanders have made changes in their behaviour in the last five years to mitigate their carbon footprints. Accordingly, climate change is an important topic in Iceland, although often discussed in economic terms. Negative impacts on fish stocks in the waters around Iceland, on nature as the main tourist attraction and increased risk of extreme weathers as well as mud avalanches are common concerns in public discourse.

Iceland is globally one of the countries with the highest share of renewable energy from geothermal and hydropower as well as the world's largest green energy producer per capita. Yet, according to EU statistics the country also generates one of the highest greenhouse gas emission levels per capita in Europe.

This means that almost 100% of the electricity produced in Iceland, including heating, is renewable and domestically produced, an advanced position in a European energy context, especially relevant considering the war in Ukraine and the debate on energy security. Regrettably, this does not necessarily imply a socially just use of renewable energy. A large share of the renewable energy produced in Iceland is low-priced electricity for international aluminium smelting companies, which are still heavy greenhouse gas emitters. Aluminium production has therefore been controversial and characterised by urban-rural tensions, between local communities, which anticipated economic benefits connected to the aluminium smelters and other groups concerned about environmental damages and health-related effects for those living close to the production.

Iceland is an associate member of the EU via the European Economic Agreement and a part of the EU Emission Trading System. The country has also committed itself to the same reduction goals as outlined in the EU's climate framework. Closer examination of Iceland's climate goals reveals that social differences are recognised to some extent in the government's climate strategy and arguably more so than in the European Green Deal. This recognition is perhaps not surprising given the fact that Iceland has long been categorised as one of the most gender equal states in the world. Furthermore, the Icelandic prime minister, Katrin Jakobsdottir, joined forces in 2019 with two other female leaders, Scotland's first minister and New Zeeland's prime minister, advocating a turn towards a 'well-being economy' driven by green thinking and social equality, instead of concentrating only on GDP and economic competitiveness.

Reykjavik, the capital of Iceland, appears to be on a similar path. As one of 100 European cities to participate in the EU's mission for climate-neutral and smart cities by 2030, Reykjavik launched its 'Green Plan' for 2030. This Plan includes three key goals of carbon-neutrality, green growth and a people-centred 'nobody left behind' goal, emphasising the importance of just transition. Accordingly, Iceland appears to have taken the first steps to become a just and inclusive carbon neutral welfare state.

Challenges and Obstacles: Heavy Industry, Transport, Fisheries and Tourism

There are though also obstacles stemming both from key industries and infrastructure. Heavy industries, particularly aluminium companies, are heavy greenhouse gas emitters, but here political and public consensus on how to proceed is limited. Transportation is another mitigation obstacle, especially since Iceland does not have a railway system and domestic flights are heavily subsidised for citizens living outside the capital area. Access to public transportation varies greatly depending on location, making inhabitants in rural areas dependent on private cars. Iceland is in general characterised by a strong car-centrism with a long tradition of private car ownership tied to most households. An important part of Iceland's move towards climate neutrality is banning the sale of new fossil-fuel vehicles from 2030 and offering tax reductions for electric cars. This action has already put Iceland in second place behind Norway for electric cars purchased per capita. These numbers can though be tied to the country's car-centrism, in other words the need to own a car in Iceland instead of relying solely on public transportation. There is also clearly a link to the country's relatively large and prosperous middle class, people who can afford to buy and own electric cars. Important also is the fact that one of Iceland's key industries, along with fishery, is tourism, which requires better climate-neutral transportation, both rental cars and public transportation.

The importance and strong interests of the fishery industry also give rise to certain challenges in transition to renewable energy, which will call for radical changes. Rural coastal communities will also be affected and might have interests which contrast with those of urban groups, primarily those living in the capital area, where two-thirds of Iceland's population are situated. Finally, Iceland is a very import-dependent state and hence the war in Ukraine has not only highlighted potential problems related to increasing energy prices, but also placed sustainability and food security higher on the political agenda.

Recommendations

As an associate member of the EU, via the European Economic Agreement, Iceland's commitments have been aligned with the Union's climate strategies. The European Economic Agreement membership, though, does not give Iceland any formal influences in EU decision-making and it is therefore important that Iceland makes the most of other means to exert its influence.

For instance, Iceland should try to develop a stronger working relationship with the European Commission and draw attention to good examples within renewable energy and the well-being economy. It is well known that the Commission often uses national examples when drafting legislative proposals and Iceland should, therefore, emphasise its reputation as an expert in renewable energy to make its voice heard in Europe, thereby striving to strengthen its position as an advocate of the well-being economy and just transition.

Domestically, Icelandic authorities should make use of the initial work already undertaken by including social differences and issues of climate justice in policy documents. An intersectional analysis would be a useful tool to tackle many of the afore-mentioned challenges. For example, this could mean a people-centred focus, instead of a traditional techno-economic analysis of heavy industries' impact on neighbouring communities and the effects of a faster green transition on small fishery communities. A similar analysis could also be further developed to decrease car-centrism, thus with a strong focus on transport patterns and access to public transportation based on location, age, gender and class.

Finally, Icelanders need to be reminded of their country's position as a European state, which not only shares mitigation responsibilities with other European states, but also has an opportunity to contribute to the development of European climate politics.

Gunnhildur Lily Magnusdottir is an Associate Professor in Political Science and Deputy Head of the Department of Global Political Studies at Malmö University. Her current research focuses on climate and energy policy-making in Europe with a specific focus on the role of governmental institutions and how they work with issues of environmental justice and social differences.

Liechtenstein: Small State, Little Responsibility?

Christian Frommelt

The extremely small state of Liechtenstein is located in the heart of the Alps, with its landscape and intact nature very much part of the country's identity. When Liechtensteiners were recently asked in a survey what their country stands for, nature was the fourth most mentioned association. Consequently, it is not surprising that when asked about Liechtenstein's most pressing problems, the environment and climate as well as energy issues are always among the five most frequently mentioned. But how important is the fight against climate change for Liechtenstein in reality?

From Vision to Strategy

Liechtenstein's politics is not known for being particularly innovative and courageous. Shaped by a political culture of consensus, but also by the lack of resources inherent in a small state, Liechtenstein prefers to wait and see which political measures have proven successful in other states before trying them out for itself. Hence, for a long time, there was no real move in matters of climate policy. However, in recent years the fight against climate change has gained some momentum. In 2020, for example, the government published a climate vision for Liechtenstein with the goal of achieving net zero greenhouse gas emissions by 2050. The current government is going even further, with sustainability being the leitmotif of its programme, Climate Vision 2050, which was followed by the concrete Climate Strategy 2050.

C. Frommelt (✉)
Liechtenstein Institute, Gamprin, Liechtenstein
e-mail: christian.frommelt@liechtenstein-institut.li

Cornerstones of the Climate Strategy

This Climate Strategy aims to achieve a 50% reduction in greenhouse gas emissions by 2030 compared to 1990, whereby 10% can be offset abroad. Measures focus primarily on the energy sector under the headings 'buildings and industry' along with 'mobility and space'. Examples of measures include an obligation to: install photovoltaic systems in new buildings and roof renovations; ban fossil heating systems in new buildings and replacements; adjust motor vehicle tax to differentiate between fossil fuel power or fossil-free public transport.

Other topics within the climate strategy are agriculture, waste and wastewater as well as the landscape in general. This third topic provides for measures such as the rewetting of moors or the greening of public areas. Indirect greenhouse emissions are also addressed in this strategy, that is those emissions that do not occur directly in Liechtenstein. Measures here are aimed in particular at raising awareness among the population and promoting climate-neutral financial investments. To the surprise of many, the government did not merely put forward its strategy, but also sent a comprehensive legislative bill for the implementation of various measures into consultation shortly after publication of the strategy.

Is There Enough Political and Public Support?

There is only one party in Liechtenstein that has long been committed to environmental concerns, the left-greens, and this party has always been in opposition, never gaining more than 15% of the vote. However, in recent years environmental issues have become popular with all parties, with none opposing increased environmental regulations. Instead, opposition against a more active climate policy is constituted across party lines. Opponents mainly refer to the liberal economic order which should do without prohibitions and strict guidelines for the economy. Accordingly, it is not surprising that measures such as the ban on fossil heating systems or the obligation to install photovoltaic systems are met with some resistance. Climate change should rather be combated primarily through technical progress and incentives in the form of subsidies. Many representatives continue to assume that there will be a trade-off between economic growth and environmental protection. In line with this view, increased environmental protection can be achieved only at the expense of reduced economic growth. However, if one takes a close look at Liechtenstein's economic structure, there are no signs of such a trade-off. Although Liechtenstein has a strong industrial sector, it does not comprise particularly environmentally intensive and damaging industries. Accordingly, there are no companies and sectors in Liechtenstein that would be particularly affected by climate policy.

Another narrative that stands in the way of an active climate policy is that of Liechtenstein's size. The country is too small to solve a global problem such as climate change and is therefore not responsible. While the first part of this statement is quite correct, the conclusion that Liechtenstein has no responsibility for the fight against climate change is very questionable.

However, with the small size there are also some real problems. For example, the possibilities of promoting renewable energy in Liechtenstein are limited *per se* and where opportunities do still exist, conflicts of goals quickly arise. In a small country, the construction of a wind farm or hydroelectric power plant very quickly represents a major intervention in the landscape. Yet paradoxically, it is precisely this love of its unique landscape that feeds Liechtenstein's support for an active climate policy. The limitations for Liechtenstein itself to promote sustainable energy production could also cause problems in the implementation of requirements by EU law. In view of this conflict, it is not surprising that the government focuses mainly on measures concerning buildings and thus mainly undertakes projects which are numerous and small rather than few and large.

Recommendations

Energy consumption per inhabitant has been slightly reduced in recent years. Likewise, the share of fossil fuels in the energy mix is declining. With a share of 34%, electricity is the most important energy source in Liechtenstein, although it is unclear from which sources the electricity originates. The share of district heating from waste has increased in recent years. Overall, the self-supply quota is only 13%.

With its most recent proposals, the Liechtenstein government has shown a surprising amount of commitment to an active climate policy. It is still unclear whether or not it will find broad support for this. It is therefore all the more important that the government tries to continue on its path consistently and does not allow itself to be distracted by disruptive fire from individual representatives of politics and business. Even the threat that a referendum will be taken against the bill and that the bill will thus be submitted to the people for approval after it has been passed by parliament should not put the government off. As the surveys mentioned at the beginning of this chapter show, the population of Liechtenstein—despite all reservations about excessive regulation—is well aware of the relevance in this fight against climate change. An exception here is, at best, the issue of mobility. Liechtenstein is a true 'car country'. According to the climate strategy, it should at least become an 'e-car country' through targeted incentives.

However, support for an active climate policy in Liechtenstein ultimately depends very much on signals from the EU. Only if ambitious and, above all, binding goals are defined at European level, will Liechtenstein commit itself to an active climate policy. This applies regardless of whether such measures are binding only for the EU states or also for the three European Economic Area/European Free Trade Association states. Ultimately, responsibility for the fight against climate change is neither a question of association status nor state size.

Christian Frommelt is Director of the Liechtenstein Institute in Gamprin-Bendern (Liechtenstein). He holds a PhD from the Swiss Federal Institute of Technology (ETH) in Zürich as well as a master's degree from the University of Innsbruck. His research focuses on the European Economic Area (EEA) and Liechtenstein's political system.

The Liechtenstein Institute was founded in 1986 as a research institute for scientific research in the fields of history, politics, law and economy relating to Liechtenstein, organised as a non-profit association and state-subsidised. It is also a member of TEPSA.

Climate Change: A Powerful Engine for Economic Transformation in Montenegro

Danijela Jacimovic and Olena Korohodova

Climate monitoring and assessments show that Montenegro has already been affected by global changes. Climate risks are rated as high to medium and include floods, landslides, wildfires, earthquakes, extreme heat and water scarcity. Moreover, the most vulnerable economic sectors that can expect to suffer the negative impacts of climate change in Montenegro include human health, tourism, agriculture, water use and other natural resource sectors. When it comes to vulnerability and the negative effects of climate change, results from observations in the period 1949–2020 show that average annual temperature values, especially over the last 20 years, have increased from 8.6 °C to 9.94 °C. Estimates of climate projections indicate that the annual temperature throughout Montenegro will increase by between 1.5 °C and 2 °C by 2040. In addition, the average annual amount of rainfall is forecast to decrease, especially during the summer months, while there is expected to be a rainfall increase over the winter months in some parts of the country, particularly towards the north.

Even though the country's national greenhouse gas emissions account for only 0.009% of the global total, Montenegro remains essentially committed to managing its development potential in a responsible and sustainable manner, in other words having minimal impact on the environment and climate change. The Paris Agreement ratified in 2017 committed to reducing greenhouse gas emissions by at least 1572 kt CO2eq, to a level of 3 667 kt CO2eq or less. Montenegro's contribution to international efforts for counteracting climate change, expressed through the Intended Nationally Determined Contribution to the reduction of greenhouse gas emissions, is set as a reduction of at least 30% by 2030 compared with the base year 1990. To achieve these goals, Montenegro has received support from the international community through various financial mechanisms, but mainly in the form of

D. Jacimovic (✉) · O. Korohodova
Faculty of Economics of University of Montenegro, Podgorica, Montenegro
e-mail: danijelaj@ac.me

loans and grants. In the period from 2014 to 2017, the state received official development assistance valued at over EUR 200 million from a number of partners, intended to be used for initiatives related to the fight against climate change. The EU has been the main source of donations, contributing about 60% of the total funds for project financing, while the United Nations and the Global Environment Facility together, through programmes and donations, contributed approximately 30% of the total funds.

The biggest impact on greenhouse gases is made by the energy sector, where production of electricity from coal—the Pljevlja Thermal Power Plant—produces the most significant contribution to carbon dioxide emissions, being 48.01% of the total in 2017. The second largest contribution to greenhouse gas emissions stems from the transport sector, followed by other sectors of industry, such as waste and agriculture. As the country's main energy resources are based on the use of hydro-power, the thermal plant plays an important role in ensuring long-term stability for the overall energy system, thereby safeguarding a reliable electricity supply, at least for the time being, bearing in mind the war in Ukraine. Fortunately, Montenegro does not use natural gas and furthermore has no infrastructure for its distribution.

Developing the renewable energy sector has enormous potential and could provide an extra boost to the country's established growth. Transition towards renewable energy could be achieved relatively quickly and easily, with the support of foreign funds. More significantly, Montenegro could not only develop a self-sufficient energy supply, but also be an exporter of green energy to the Single Energy Market. At the same time, the energy sector might be an important means of diversifying the country's economy and providing significant support to the tourism sector by expanding the range of growth sectors.

Significantly, the production of green energy could also be a key factor in balancing regional development in Montenegro. It is evident that the economic structure is divided across three very unbalanced regions according to 2019 data from the Central Bank of Montenegro and the authors' own calculations. The country's northern region represents only about 5% of Montenegro's economic activity, although it makes up more than half of its territory, while it is home to less than a third of the total population and has only 50.1% of the country's average development level. The central region (where the capital Podgorica is situated) represents 60% of overall economic activity, makes up 35% of the territory of Montenegro, but is home to 50% of the population. The southern region produces 35% of the economic activity, makes up only 11% of the territory and is home to 25% of the total population.

The most significant hydro potential in Montenegro is situated in the northern region, where building new hydro plants could balance economic development, provide a stable electricity supply and encourage establishment of enhanced and necessary transport infrastructure. All of that will lead to a wider range of sustainable tourism in the country, would include the northern region with its tourism and agriculture potential and most importantly potentially limit or even reverse its depopulation.

Windmills could be built in the central region, while the most significant potential for solar plants is found in the south. Based on analysis from global solar radiation maps for Montenegro, it was concluded that Montenegro has great potential for the use of solar energy. There is over 2000 h of sunshine per year for most of the territory, while for the region along the coast, it is over 2500 h per year. Assuming that the average solar insolation in Montenegro is 1450 kWh/m^2 per year, the potential for solar radiation could be estimated at 20 PWh/year. Solar plants would greatly diversify the southern region's economy, provide the necessary electricity supply to the tourism sector and balance the electricity system in summer, when production of hydropower energy is low. In this way, such a scheme would contribute to reducing reliance on the thermal plant and its importance to the electricity system's stability, enabling coal power to be reduced and eventually phased out entirely.

Recommendations

It seems logical for Montenegro to build up resilience to climate and disaster risks proactively, whilst at the same time ensuring its own energy security in order to boost the country's potential growth. This would significantly contribute to creating a more balanced form of regional development. Some specific recommendations include the following: (i) European Investment Fund resources should be used to stimulate a fast, green and effective transformation of the energy sector, particularly in terms of financing large regional hydro plants and expensive accompanying infrastructure in the country's northern region, so as to speed up its implementation; (ii) Policy-makers should be aware of the momentum that has built up to support a green energy transition and offer favourable investment as well as business conditions to attract European private investors (for example to finance solar energy projects); (iii) The promotion and adoption of innovative technologies through a specific technology transfer mechanism is another important channel for the green transition of the economy and society, alongside significant support from our integrational partners; (iv) Raising public awareness and education in the area of green energy transition and its potential economic benefits is vital. According to a UNICEF survey, slightly more than a third of Montenegrins say that they often or regularly follow the news on climate change, while a quarter never or rarely do so. More outreach on the topic would therefore encourage support for the green energy transition in both economic and social terms.

Danijela Jacimovic is a Professor at the Faculty of Economics of the University of Montenegro. Her fields of interest include International Economics and European Integration.

The University of Montenegro is a public higher education institution and as such is the oldest in Montenegro. The Faculty of Economics, as one of the most important educational and research institutions in the country, is also a member of TEPSA.

Olena Korohodova is an Associate Professor at Igor Sikorsky Kyiv Polytechnic Institute and a current Visiting Professor at the Faculty of Economics of the University of Montenegro. Her fields of interest include International Economics.

The University of Montenegro is a public higher education institution and as such is the oldest in Montenegro. The Faculty of Economics, as one of the most important educational and research institutions in the country, is also a member of TEPSA.

North Macedonia: Pause on the 'Green Agenda' During Crisis

Irena Rajchinovska Pandeva

Heightening Public Awareness

North Macedonia is a small non-coastal country in the Western Balkan region with a diverse climate that ranges from alpine to Mediterranean, reflected in cold winters and very hot summers together with a tremendously variable precipitation regime. The whole region is deemed to be one of the most affected by climate change, within which North Macedonia is no exception. It is infamous for being one of the most polluted as well as one of the poorest regions in Europe, hence, 'going green' is more than just a financial and structural endeavour, it is a question of sustainability, development and improvement of living conditions as well as the general quality of life.

Inadequacies in waste management systems, wastewater treatment and water quality, combined with high air pollution and issues concerned with liberalisation of the energy market as well as many other serious concerns, paint a very grey picture of this green country. For instance, the air pollution in 2018 and 2019 surpassed the ceiling for air pollutants determined by the national emission reductions plans (sulphur dioxide and dust being particularly problematic), which resulted in the opening of a case against North Macedonia (and Kosovo) by the Vienna-based Energy Community Secretariat. According to national polling data, public opinion ranks the lack of clean water as the biggest social problem (60%) and despite the effects of the global COVID-19 pandemic, climate change (59%) was considered to be a more serious threat than infectious diseases in 2021.

Data shows that apart from household heating and traffic, the thermal power plants themselves are mostly responsible for the country's air pollution. According to national research studies and a 2019 World Bank report, air pollution is a grave problem that has health implications and hence economic costs associated with

I. R. Pandeva (✉)
Facutly of Law Iustinianus Primus, Ss. Cyril and Methodius University, Skopje, North Macedonia

© The Author(s), under exclusive license to Springer Nature Switzerland AG 2023   151
M. Kaeding et al. (eds.), *Climate Change and the Future of Europe*, The Future of Europe, https://doi.org/10.1007/978-3-031-23328-9_34

health impacts. Additional data from the Balkan United for Clean Air campaign indicates that 3400 deaths per year result from exposure to microscopic particulate matter ($PM_{2.5}$) fine dust particles, responsible for a 10% reduction in fertility rates and a 13% increase in the risk of miscarriages and stillbirths. A recent account on North Macedonia as 'Europe's Waste Dump' by investigative journalists was linked *inter alia* to corruption, the country's weak institutional setting, a lack of responsibility and poor law enforcement.

At the turn of the millennium, North Macedonia was praised for reducing emissions related to land-use change and forestry, which according to research data amounted to an 8% fall in greenhouse gas emissions mainly due to agricultural production. It was also the first country in the Western Balkans to decide on abandoning coal. The ecological turnaround of the Western Balkans and North Macedonia was more recently encouraged by the EU's Green Agenda (signed by six states in Sofia at the 2020 Western Balkans Summit) to support countries in establishing their plans for climate neutrality by 2050. The Green Agenda was envisioned as a roadmap for dealing with climate change predicaments by building upon new energy and mobility solutions, sustainable economies, environmentally friendly agriculture and preservation of biodiversity. A complementary path was set in 2006 when North Macedonia and six Western Balkans countries joined the Energy community.

Pushing the Green Agenda for the Western Balkans was not simply a matter of exporting the EU Green Deal on a regional level. Essentially, not only did the Deal have to be linked with the region's advancement as part of the EU integration process, but it also needed to be in line with commitments undertaken by governments in international settings. North Macedonia is a non-Annex I party to the United Nations Framework Convention on Climate Change and ratified the Paris Agreement in November 2017, following which the country introduced an Energy Law in May 2018. This new Law facilitated reforms in the energy sector to support renewables investment projects and harmonised North Macedonia's energy legislation with the EU's Third Energy Package. The long-awaited national Energy Development Strategy which took some years to prepare was adopted only in early 2020.

The latest energy crisis stretched to 2022 so North Macedonia was compelled to extend the state of energy crisis which was declared on 9 November 2021. In the early spring of 2022, a state-owned electricity production company Elektrani na Severna Makedonija announced the opening of two lignite mines, this retrograde step being justified as a necessary reaction to the crisis bearing in mind the comparative advantage of coal as a cheap and secure energy source. As of July 2022, the country's daily electricity production amounts to 86% by lignite-fired power plants, 8% by hydro power plants and 6% by windmills.

Prior to the current energy crisis, North Macedonia announced that EUR 8.2 billion will be invested by 2027 (EUR 3.1 billion in energy) according to the Intervention Investment Plan, including various projects covering: solar and wind power; waste separation and management; sewage and waterworks; wastewater processing; along with railways and gas infrastructure. The Government also announced that it will seek external funding to support its Plan and reported that negotiations with a German investor are being finalised on construction of a wind farm that will cost around EUR 500 million.

In recent years the country has increased investment in its gas distribution infrastructure aiming to use gas as an additional but most importantly transitional energy source to support the coal and lignite phase-out. Until recently, gas supplies came from Russia through Bulgaria, but its nationwide use (especially by households) is still very limited, even though its capacity was increased in 2021 by 26% in comparison to 2020. Optimistically North Macedonia will manage the supply in the long run since it has invested in a project of gas interconnection with Greece which is planned to be operational by 2024. However, the handling of any immediate shortage remains uncertain at this point.

The current energy crisis has highlighted two significant flaws in the state's strategy and management of the energy sector: the question of development and investment, together with the issue of almost complete dependency on coal. Both factors have led the country not only to a serious energy crisis but also a great financial burden for its population. To ease the effects of this energy crisis, the government initially spent about EUR 120 million, began a procedure to amend the rulebook regulating the maximum capacities of rooftop solar power plants and made efforts to maintain the transition of its economy from fossil fuel dominated to energy production from solar and wind power, hydropower and gas. The national plan envisaged a coal phase-out and switch to renewables within a decade and achieving climate neutrality by 2040. However, in light of the crisis, the government has been forced to postpone its coal exit from 2027 to 2030. Furthermore, crisis mitigation action has resulted in emergency purchases of coal for two power plants and oil for another. Bearing in mind that the two coal power plants produce 65% of the country's electricity, this step was necessary albeit unwanted.

The latest investments such as the wind farms in Bogdanci and Miravci as well as the Oslomej photovoltaic park do bring hope that policies can lead to action, but for projects to be realised state investment is imperative. Sadly, records reveal that the total energy produced by renewables in 2021 amounted to only one-third of that produced by the biggest coal power plant.

Recommendations

Energy security will be on top of government priorities in the upcoming period and as announced, in the short-term solutions will first be pursued within national capacities by mining and using coal at the expense of temporarily abandoning the green agenda. At the same time, efforts to increase renewable energy production must continue, on the basis of a just and inclusive transition. Of course, progress is related to the clearing of North Macedonia's EU integration path, which will without a doubt have a very strong and direct impact on the country's climate and energy policies as well as its actions over the course of the next decade. Despite public awareness of climate change impacts and the government's proclaimed willingness to moderate its effects by promoting solutions and policies, ultimately this cannot be managed without external support and guidance.

Irena Rajchinovska Pandeva has a background in political science and international relations. She works as professor at the Iustinianus Primus Law Faculty at the Ss. Cyril and Methodius University in Skopje, where she teaches several courses at undergraduate, master and doctoral levels. She is also vice dean for science and international cooperation of the faculty. She is alumni of Fulbright (US Institutes), CEEPUS, ERASMUS and OEAD programmes, local coordinator of the CEEPUS network 'Ethics and politics in European context', director of the Refugee Law and Migration Summer school at UKIM and member of the TEPSA Board.

The Iustinianus Primus Law Faculty, Skopje is the oldest institution of higher education in the country, offering legal studies programmes and is part of the Ss. Cyril and Methodius University in Skopje and the first public University in North Macedonia. It is also a member of TEPSA.

Norway's Climate Policy: Don't Think of the Elephant!

Kacper Szulecki

All Norwegian governments in the twenty-first century, left and right, have made climate action an important element of their diplomacy and domestic policy, while recently some political parties have even made climate neutrality and decarbonisation the core of their electoral campaign messages. Norway has been an international promoter, for instance of rainforest protection. Moreover, government incentives such as tax levies have been instrumental in the spectacular expansion of electric vehicles. However, despite the self-promoted image of a climate policy champion abroad, Norway's efforts to cut domestic greenhouse gas emissions have been modest since signing the Kyoto Protocol in December 1997. As noted in a recent editorial from the country's leading daily newspaper, Dagbladet, over the last three decades, Norway has 'cut 3.2% in emissions. Over the next three decades, we have to cut just about everything. It will affect all areas of society. Nevertheless, there is remarkably little talk about this so far [. . .] It almost seems as if many politicians are terrified'.

Norway's perceived failure to achieve emission cuts has been criticised both domestically and internationally, most recently by an OECD report, but this apparent lack of progress is easy to explain. Unlike most European countries, Norway cannot register any quick wins, as it already boasts a near-carbon-neutral electricity system, based on clean hydropower. For that reason, domestic political debate and climate policy efforts have been focused primarily on the decarbonisation of transport, as well as mitigation and adaptation efforts in the global South, such as the Reducing Emissions from Deforestation and Forest Degradation programme. If the oil and gas sector's carbon footprint has been considered, it was only in terms of its direct emissions, meaning those occurring in the process of exploration and production. Hence, carbon capture and storage together with the electrification of platforms on

K. Szulecki (✉)
Norwegian Institute of International Affairs (NUPI), Oslo, Norway
e-mail: kacper.szulecki@nupi.no

© The Author(s), under exclusive license to Springer Nature Switzerland AG 2023   155
M. Kaeding et al. (eds.), *Climate Change and the Future of Europe*, The Future of
Europe, https://doi.org/10.1007/978-3-031-23328-9_35

the Norwegian continental shelf (making them emissions-free by powering them with electricity from offshore windfarms or onshore renewables) have been prominent policy topics.

This leads to an obvious tension, opening Norway up to not unfounded accusations of hypocrisy. The elephant in the room, which Norwegian policymakers prefer to ignore, is the oil and gas sector, a source of national wealth and a pillar of the robust welfare state. It represents 12% of gross domestic product and 36% of total Norwegian exports, employing—directly and indirectly—some 6% of its workforce. Having said that, this sector is also Norway's largest contributor to the global climate crisis. Estimates put greenhouse gas emissions from Norwegian produced oil and gas at around 500 megatons of carbon dioxide per year, a figure that puts Norway in the global top ten worst offenders. However, since the United Nations Framework Convention on Climate Change regime focuses on emissions 'from the exhaust pipe', the elephant has remained ignored for many years.

Following the 2015 Paris Agreement, the 2018 emergence of new protest movements and the proliferation of climate crisis rhetoric reveal signs of change. Over the past decade, the notion of an end to oil and gas production in Norway was in the realm of political fiction, as with the popular TV series 'Occupied', which depicts a Green Party landslide and the new prime minister's decision to stop all petroleum production with immediate effect. Recently, proposals for a phase-out of domestic oil and gas production were tabled by a number of political parties in the run up to the 2021 elections. While the victorious coalition of Labour and agrarian-populist Centre does not as yet envisage an end date for oil and gas production in Norway, recent reports from the Intergovernmental Panel on Climate Change, the need to fulfil Paris Agreement's obligations, helped to open a new debate on green transition and carbon neutrality. This would see if not a complete phase out then at least a significant running down of production on the continental shelf. An additional factor is the pressure from new ambitious EU climate targets, to which Norway, although a non-member, is usually trying to adapt. EU policies also have an indirect impact through energy and climate policy regulation (the Clean Energy Package and the recent Fit for 55 package); the functioning of the internal market in which Norway is a member via the European Free Trade Association/European Economic Area agreement; the EU Emissions Trading System, which also covers Norway.

However, in the wake of Russia's invasion of Ukraine and the EU's decision for a drastic reduction in dependence on Russian fossil fuels, the lifetime of Norway's oil and gas production could well be extended by a decade or more. Longer term, it is not unreasonable to expect that the last molecule of fossil methane burned in Europe before it switches to hydrogen and biogas—is going to come from Norway.

Notwithstanding the geopolitically driven delay, Norway is set for an ambitious green transition. It will affect the oil and gas industry directly and consequently the entire economy. Some 140,000 to 200,000 people are employed in this sector as well as related industries and thus securing alternative jobs for them will be a major issue. Norway's longstanding tradition of tripartite negotiations makes trade unions, to which over a half of the workforce belongs, key players in the transition. Unions' views on the transition vary, depending on the skills, education and perceived flexibility of their members, but main labour confederations have been active in constructive debates on decarbonisation. As with the Labour party, they are under pressure from their youth organisations, who are anxious to talk about the elephant in the room.

Significantly, a large part of Norwegian society remains sceptical about climate change science and the need for ambitious and costly climate action. In a 2019 international survey, Norwegians ranked among the most climate sceptical, with only 35% of respondents ascribing climate change to human causes. Since 2014, strong grassroots protest movements have emerged around the issues of environmental road tolls as well as onshore wind energy and high-voltage electricity cables. The Agency for the Cooperation of Energy Regulators has become an unlikely symbol of Norway's asymmetrical relationship with the EU, due to the European Economic Area agreement, making the Agency's acronym - ACER - shorthand for 'Brussels' dictates trumping Norwegian sovereignty' for populist politicians on the right (Progress Party), the left (the Reds) and in the Centre party. The unprecedented electricity price hike in 2021 and 2022, for which some politicians blamed climate policy as well as the EU market, shows that future political debates will be contentious. The EU's increasing ambition, the broad European Green Deal strategy and the Fit for 55 package will increasingly make climate policy and views of European integration salient political cleavages.

Whilst Norway's point of departure in an imminent transition is rather favourable, the lack of progress is due to insufficient political leadership and vision. Norwegian decision makers need to be bold in their choice of whether the green transition's main goal should be managing decline in the oil and gas sector or managing climate-related economic risks.

The offshore wind industry, carbon capture and storage, green hydrogen and a modern bioeconomy have all been identified as potential alternatives for a robust low-carbon economy, but progress in these areas is slow. Norway needs concrete strategies for these sectors, coupled with strong and binding decarbonisation targets, in line with the Paris Agreement.

Recommendations

For political reasons, Norwegian oil and gas will continue to fuel the EU in the foreseeable future. However, over time exporting hydrogen combined with importing CO_2 for sequestration can assist in the EU's green transition. The expansion of cross-border and undersea cables together with exporting flexible peak energy from Norwegian hydro plants can already help Europe both in its decarbonisation efforts and to some extent in the short-term crisis resulting from Russian supply interruptions. Most importantly, exports of electricity can constitute a much quicker and more substantial contribution by Norway to European climate efforts than future technologies such as hydrogen as well as carbon capture and storage. Whilst this may be politically costly, building Norwegian welfare on exporting carbon while egoistically safeguarding cheap green power at home is ethically unsustainable.

Norway's integration with the EU so far as the European Green Deal is concerned will need to be even greater. Norwegians should use their privileged position (as suppliers of energy, technology and important raw materials) not only to shape the transition process, but also influence regulation in a manner that safeguards vital economic interests, most importantly Norway's full participation in the common market with strategic goods, such as batteries, without locking-in environmentally harmful practices. The EU-Norway green partnership, announced only a day before Russia's invasion of Ukraine, will gain in significance and can become a platform through which Norwegian authorities can constructively upgrade this trade, energy and political interdependence.

Kacper Szulecki is a research professor in the Climate and Energy Research Group at the Norwegian Institute of International Affairs (NUPI), author and editor of numerous publications on energy and climate policy, such as 'Energy Security in Europe' (Palgrave 2018). NUPI is a leading institution for research and communication about international affairs, currently ranked 47th among non-US think tanks, making it Norway's top and one of Europe's leading research institutes.

It is a major centre for research on international issues in areas of particular relevance to Norwegian foreign policy. As such, the Institute communicates research-based insights to the Norwegian public as well as wider international audiences, being committed to excellence, relevance and credibility. NUPI is also a member of TEPSA.

Switzerland's Climate Policy: Caught Between EU-Compatible Goals and Referendum Constraints

Frank Schimmelfennig

As an Alpine country, Switzerland is particularly strongly affected by climate change. The progressive glacial melt is testament to the effects of global warming. Temperatures are rising twice as fast as the global average. In 2019, the Federal Council of Switzerland decided that Switzerland would become 'climate neutral' by 2050. Moreover, the government is committed to cutting greenhouse gas emissions in half by 2030 (based on 1990 levels). The Swiss goals are thus highly compatible with the EU's Green Deal targets. In other ways, Swiss climate policy also tracks that of the EU. For instance, Swiss and EU emissions trade is linked, whilst Switzerland and the EU share emission reduction targets for cars.

However, the Climate Action Tracker rates Swiss policy as 'insufficient'. Whereas its domestic targets come close to meeting the goal of limiting global warming to 2 °C above the pre-industrial average, its current policies would not. Switzerland's international public finance contributions are even considered 'highly insufficient' and far away from its fair share of targets. In considering the European comparison, it is remarkable that almost half of Swiss CO_2 emissions are related to road transport and aviation, despite a dense and efficient public transport system. Industry accounts for roughly one-fourth and agriculture one-sixth of the country's greenhouse gas emissions.

The current discrepancy between Switzerland's international commitments and actual policy results from the revised CO_2 Law having been defeated in a June 2021 referendum. This law envisaged enhanced measures to support energy transformation, including a higher levy on fossil fuels and a new levy on air tickets. Whereas the law had enjoyed broad parliamentary support, it was rejected by the populist far-right Swiss People's Party, Switzerland's largest political grouping. Moreover,

F. Schimmelfennig (✉)
ETH Zurich, Zurich, Switzerland
e-mail: frank.schimmelfennig@eup.gess.ethz.ch

this referendum against the law was supported not only by the car, fuel and transport sector associations, but also 51.6% of Swiss voters.

The referendum showed that while an ambitious climate policy enjoys abstract support, a narrow popular majority may still reject concrete restrictions to the freedom of consumption, even in a country as strongly affected and prosperous as Switzerland with its comparatively liberal policy measures. Switzerland's direct-democratic mechanisms are particularly prone to revealing this tension. Even though the Swiss government continues to adhere to agreed medium and long-term goals, the electoral defeat of the CO2 Law has cast severe doubt on Switzerland's actual ability to meet its climate policy objectives and international obligations. The Swiss government is currently working on a revision of the law, putting stronger emphasis on support for climate-friendly investments, forgoing new levies and offering more generous exemptions than those which already exist so as to placate opponents. The new law continues to be rejected by the Swiss People's Party, though. It may therefore be challenged in yet another referendum.

At the same time, a popular initiative by supporters of a more ambitious climate change policy—the so-called 'glacier initiative'—aims to anchor the 2050 net-zero goal in Switzerland's constitution. It also strives to prohibit fossil fuels altogether and rejects giving firms credit for emission cuts abroad. By contrast, the government supports a parliamentary counterproposal to achieve the net-zero goal through future revisions of the CO2 Law and without the proposed prohibitions. Even if this CO2 Law survives the next referendum, it remains questionable as to whether or not a liberal climate policy renouncing bans and levies will be effective enough and generate sufficient resources for the necessary investments in energy transformation.

In any event, Swiss climate goals require a far-reaching transformation from fossil fuels to electricity. At the same time, Switzerland plans to replace its nuclear power stations, which currently provide a third of the electricity consumed in the country, with renewable sources. Electricity from renewables will thus become the key source of energy. In this regard, there is a clear link between Swiss and EU policy. For a small country in the middle of Europe, security of supply requires full integration with the European grid to compensate for the fluctuating availability of hydropower, wind power and solar energy. Switzerland has 41 electricity connections to its neighbours (more than any other country in the world); it is an important transit country for electricity (especially from Germany and to Italy), and it used to be a crucial partner in the development of the Europe-wide electricity grid. However, Switzerland was side-lined by the integration of EU electricity policy, excluded from participation in the relevant agencies and cut off from the coordination of load flows through the country.

Switzerland's free access to and participation in coordination of the EU electricity market is the cornerstone of the bilateral Swiss-EU electricity agreement, under negotiation since 2007. This electricity agreement is in the interests of both the EU and Switzerland. As such it is uncontroversial. Yet, the EU has linked its conclusion to the Institutional Framework Agreement, which envisages *inter alia* the dynamic adaptation of Swiss legislation to EU law in areas of market access as well as the introduction of an arbitration court. In May 2021, though, the Swiss government broke off negotiations unilaterally. Since then, Swiss market access, for instance to do with medical technology products, has eroded and negotiations on further economic integration, as in the electricity market, have stalled.

In addition, the Russian invasion of Ukraine has a significant impact on Swiss energy policy. Even though in comparative terms Switzerland is only weakly dependent on Russian gas (which accounts for 47% of the gas imports, but only 7% of the Swiss energy consumption), all of it is delivered through EU neighbours as the country does not have significant storage capacity of its own. Nor is Switzerland a party to the EU solidarity arrangements for gas supply.

Recommendations

Both longer-term requirements of climate change and short-term pressures due to the Ukraine war demonstrate EU and Swiss interdependencies in energy transformation. Switzerland and the EU should continue to develop their climate policies in parallel and achieve climate neutrality together by 2050. Moreover, to reduce costs and secure supply, Switzerland needs to be fully integrated into the EU energy markets. The current energy crisis and the high risk of electricity shortages in the future are the most important reasons for unblocking negotiations on the Institutional Framework Agreement. The Swiss government should not waste this opportunity.

Frank Schimmelfennig   is a Professor of European Politics at ETH Zurich. He is also a member of the Swiss National Research Council, an Associate of the Robert Schuman Centre for Advanced Studies at the European University Institute, Chairman of the Scientific Board of Institut für Europäische Politik Berlin and a member of the Board of the Trans-European Policy Studies Association (TEPSA). His research focuses on the theory and development of European integration. His most recent books on the EU are Ever Looser Union? Differentiated European Integration (Oxford University Press, 2020, with Thomas Winzen) and Integration and Differentiation in the European Union: Theory and Policies (Palgrave, 2021, with Dirk Leuffen and Berthold Rittberger).

The Centre for Comparative and International Studies (CIS) is a joint research centre of political scientists at the ETH Zurich and the University of Zurich, offering a joint master's programme (MACIS). CIS is also a TEPSA member.

Türkiye: A Climate Financing Opportunist?

Ezgi Ediboğlu Sakowsky

Türkiye finally ratified the Paris Agreement in 2021, declaring that Türkiye 'will implement the Paris Agreement as a developing country', thereby continuing its ongoing battle to be recognised as such under the United Nations climate change regime, a development which would prompt considerable climate financing. It was probably no coincidence that Türkiye ratified this Agreement after the World Bank, Germany and France had promised some additional bilateral climate financing. Following an announcement about the European Green Deal, especially its Fit for 55 legislative package pursuing climate goals, Türkiye finally realised that it must eventually transition into a low carbon economy. Moreover, climate financing could be a means to this end.

In 2021, Türkiye also announced its 2053 net-zero emissions target and is revising most of its climate change-related policies and regulations, including the nationally determined contribution. However, as of early August 2022, existing plans do not indicate a clear path to formal confirmation of the country's targets. In terms of Türkiye's objectives, it is necessary to consider two alternatives. On the one hand, the country could be seeking funding to aid its economy by adopting climate change-friendly language. On the other hand, it could be genuinely aiming at receiving funding for combatting climate change. Considering that climate change-related activities have amounted to a very low percentage of public spending so far, a legitimate concern for Türkiye's efforts emerges: is the country merely a climate finance opportunist?

E. Ediboğlu Sakowsky (✉)
Istanbul Policy Center, Sabancı University-Stiftung Mercator Initiative, Istanbul, Turkey

Max Planck Institute for Innovation and Competition, Munich, Germany

Factors Influencing Türkiye's Attitude

Drawing any clear conclusion about Türkiye's motives for seeking climate financing is impossible for the moment. Whilst at the time of writing, revised climate change plans and policies have yet to be published, the following factors are likely to influence the country's attitude.

Türkiye's extraordinarily high inflation rate and its troubled economy create incentives for its leaders to view the fight against climate change primarily as a funding opportunity. Officially the year-on-year inflation rate for July 2022 was 79.60%, while non-governmental independent sources rate it as closer to 150% or even higher. Russia's military invasion of Ukraine has also raised the stakes in this regard since Türkiye is dependent on Russian gas, which is a critical source for the country as its energy mix is still heavily carbon-based. Gas, oil and coal were identified as constituting approximately 25% to 30% each of the Turkish energy supply in 2019. Russia has been Türkiye's main gas supplier since the early 2000s and even though the country started to import gas from Azerbaijan and Iran as well as other countries, the Russian share has never fallen below 30% of total supplies.

Increased energy prices dealt a severe blow to most of the country's political and economic actors. Moreover, the energy crisis became the final straw for Türkiye's ongoing issues about high inflation, unemployment, immigration and border security. This is also because Türkiye prioritised energy, building, industry, transport, waste, agriculture, land use and forestry sectors in its existing climate change policies, with the energy sector being the key sector for catalysing any decarbonisation efforts. The Russian-Ukrainian war has placed great strain on Turkish decarbonisation plans and as a result the government has been forced to prioritise security and economy-related issues over its environmental ambitions. Even though the younger generation in Türkiye is more conscious of looming environmental dangers, most citizens would probably still rate security and the economy as more important.

It should be noted that there are examples of increasingly strong local environmental groups positioning themselves against local mines, hydroelectric facilities and fossil fuel plants. The widespread forest fires in 2021 have also raised awareness amongst the general public of the need for environmental protection and brought some non-governmental environmental organisations to the forefront, key amongst which is the Turkish Foundation for Combating Soil Erosion, for Reforestation and the Protection of Natural Habitats. Nevertheless, there are still only a few strong nationwide public figures, movements and organisations specifically targeting the issue of climate change.

However, from a business perspective, Turkish industry continues to work more carefully on realising its decarbonisation plans, since its largest trading partner is the EU, which will impose environmental standards and a carbon border adjustment on Turkish exports in accordance with the Green Deal. Hence, unlike issues related to security and the economy, this Green Deal forces the Turkish Government to improve environmental standards, thus supporting its industry in making Turkish standards compatible with those of the EU. For instance, Türkiye is starting to develop a domestic emissions trading system. The Green Deal is one of the key factors that will not only improve Türkiye's climate change efforts amidst its serious social, political and economic hardships, but also exert pressure on sectors with higher emission levels. The energy sector alone was responsible for over 70% of Turkish greenhouse gas emissions in 2019, followed by agriculture (around 13%), and industrial processes and product use (around 11%). Whilst new regulations and project financing focus on all these sectors, clearly the energy sector will be prioritised.

Recommendations

Türkiye is currently at a crossroads in planning concrete climate change targets. There are diverse factors affecting the next steps: economy and security-related issues; the Russian-Ukrainian war; energy crises; and the Green Deal. In these circumstances, there is a chance, therefore, that Türkiye will approach the climate change issue as an economic opportunity. Hence, recommendations which follow are put forward with the aim of avoiding this risk.

Firstly, Türkiye should establish itself with its renewed targets and actions as a serious actor in the fight against climate change.

Secondly, it should keep focusing on the energy sector as key in the fight to achieve decarbonisation. The global energy crisis should be a lesson for all countries to quit carbon-based energies not only for climate-related reasons but also for energy security and independence. This should be achieved by rationalising economic policies.

Thirdly, Türkiye should finalise its adjustment plans for the Green Deal without further delay (especially with regard to the domestic emission trading system) and create a better environmental standardisation for industry activities.

Fourthly and finally, public awareness of climate change implications must be boosted.

The EU could support Türkiye in this process by allocating project-based funding and technical support from its development banks. The European Bank for Reconstruction and Development, for example, has been active in this regard since 2010 and has created a number of financing facilities for many sectors, such as the Turkey Sustainable Energy Financing Facility; the Mid-size Sustainable Energy Financing Facility; the Turkish Residential Energy Efficiency Financing Facility. This kind of support should target not only governmental actors, but also domestic authorities, small enterprises, companies and universities (for instance, via Horizon Europe). The EU should also support Türkiye with its transfer of environmentally sound technologies, an obligation the EU has incurred under the UN climate change regime treaties.

Considering all of the above, is Türkiye a climate financing opportunist? We do not know until the new plans are revealed, but it seems likely. Türkiye can avoid this label only by allocating most of the climate finance funding obtained to projects that will significantly reduce emissions, rather than using it as a tool for economic capacity building.

Ezgi Ediboğlu Sakowsky is a legal scholar, holding both LLM and PhD degrees from the University of Aberdeen. Her work concerns the UN climate change regime, international institutions and the transfer of environmentally sound technologies. She is a 2021/22 Mercator – Istanbul Policy Centre (IPC) Fellow. The Mercator–IPC Fellowship Programme provides international scholars with the means for undertaking academic and practical projects at IPC, a global policy research institution.

Founded in 2001, the Istanbul Policy Centre (IPC) is a global policy research institution that specialises in key social and political issues ranging from democratisation to climate change, transatlantic relations to conflict resolution and mediation. The IPC offers policymakers, academics and young researchers a unique platform where sound academic research in social sciences shapes hands-on policy work. It is also an associate member of TEPSA.

United Kingdom: BREXIT, Climate Change and the Conservative Party

Brendan Donnelly

Since 2010 the Conservative Party has been the leading force of British politics. Its internal debates and controversies have shaped all British political decision-making throughout the past decade. Environmental policy is no exception to this rule. Just before the United Nations Climate Change Conference 26 summit in Glasgow last year, 75% of British adults were reported as feeling anxious about climate change. The Conservative government's environmental policies are though largely the result of incestuous internal debates and controversies within the Conservative Party, a discussion shaped over recent years by the Brexit referendum and its consequences.

To all outward appearances, Conservative governments have generally been content in the last decade to remain within the European mainstream on the issue of climate change. The recent British chairmanship of COP26 was widely praised and the government was proud to announce to the conference that more than half of the UK's largest businesses have undertaken to eliminate their contribution to carbon emissions by 2050. The government also makes funds available for smaller businesses, often hesitant in the UK about the impact of environmental legislation on their business models, to 'go green'. In 2019 Theresa May's government passed legislation committing to a legally binding target of net zero carbon emissions by 2050.

However, the Conservative Party's leaders have always been uncomfortably aware that there exists within their party an important minority of sceptics, even deniers of man-made climate change. Such publicly dissident voices have always primarily been found on the right of Britain's political spectrum. This vocal minority saw its influence enhanced by a significant overlap before 2016 between its members and those of the anti-EU groups that were preparing the ground for Brexit. It was a frequent trope of the Leave campaign in 2016 that British business needed to be

B. Donnelly (✉)
The Federal Trust, London, UK
e-mail: brendan.donnelly@fedtrust.co.uk

liberated from stifling European legislation, including in particular excessive environmental legislation.

Almost the whole focus of British politics since 2016 has been an attempt to work out and impose a general narrative about just what it was that the British people had voted for in that year. Environmental issues have been one battleground over which this debate has ranged. Earlier in the year, the Net Zero Scrutiny Group was set up in the British Parliament, a group calling for cuts in green taxes and an increase for fossil fuel production. One of the Group's founding members, Steve Baker, was a prominent member of the European Research Group, the pro-Brexit group which rebelled so repeatedly and successfully against Theresa May. Nigel Farage has said that at the next General Election, he will be calling for a referendum on the government's Net Zero target for 2050. This demand will certainly find a sympathetic response from many members of the Conservative Party, who tend anyway to be suspicious of what they regard as the 'political correctness' of environmental concerns and campaigning.

At the time of writing, the Net Zero Scrutiny Group is a minority within the Conservative Party. Boris Johnson was believed to favour interventionist policies to limit climate change, perhaps encouraged by his wife, who has described herself as a 'climate activist'. However, there is no guarantee that his successor Liz Truss will maintain this policy. The Trade and Cooperation Agreement between the UK and EU contains non-regression clauses on environmental standards, which should prevent dismantlement of the European environmental standards currently observed by the UK. It is possibly reasonable to hope that climate change will not be a major source of contention between the EU and the UK.

Even so, some caution is justified. The UK imports only 8% of its energy needs from Russia, the remainder being met by reliable suppliers elsewhere. But already the argument is being made by some in the Net Zero Scrutiny Group that the general economic disruption caused by the war in Ukraine should give the British government some pause for thought before adopting environmental policies that may increase the cost of living for the poorest. There are here clear echoes of the 2016 rhetoric, which claimed that British membership of the EU was a project of metropolitan elites who ignored the baneful consequences of this membership for the most vulnerable in society. As poorer British families struggle to meet their energy and other bills, commentators and pundits will be eager to argue that middle-class obsessions with climate change should not be given priority over pressing present economic needs.

Recommendations

With a strong lead from the Conservative leadership, climate change could be an area in which the EU and UK might work together over the coming years relatively harmoniously. If (by no means a certainty) President Macron's European Political Community is a success, it might well embrace climate change as one of its components. The argument made by some Leave campaigners in 2016 that

European environmental legislation was an unacceptable restraint upon the British economy maintains its traction within certain elements of the Conservative Party. But it is not popular among the wider British public. If Rishi Sunak wishes to follow his predecessors' on the whole good example in regard to climate change, he must urgently: reassert the UK's full-hearted commitment to the Net Zero target; improve the means of monitoring progress towards this target; and stress publicly that the Green Agenda is not an imposition by one section of society on another, but rather a project from which all can benefit. Until he has adopted a clear approach on this subject, the UK will always be an uncertain partner for the EU on the issue of climate change, as it is on many others.

Brendan Donnelly has been Director of the research institute The Federal Trust in London since January 2003. He was a Conservative MEP from 1994 until 1999 and had previously worked for the Foreign Office, the European Commission and the Conservative Group in the European Parliament.

The Federal Trust is a research institute studying the interactions between regional, national, European and global levels of government. Founded in 1945 on the initiative of Sir William Beveridge, it has long made a powerful contribution to the study of federalism and federal systems. It has always had a particular interest in the European Union and Britain's place in it. The Federal Trust is also a member of TEPSA.

Ukraine: Revisit Climate Goals After the War to Increase Ambition

Svitlana Chekunova

The Russian invasion has affected lives and livelihoods, not only disrupting supply chains in Ukraine and the world, but also threatening energy security and food availability. It has made climate mitigation and adaptation processes even more complicated due to the war having inflicted losses to infrastructure and ecosystems as well as aggravating economic conditions and causing a humanitarian crisis. Whilst in the short-term green transition is now facing a number of additional challenges which will slow down the process, it is likely to be accelerated in the long run. Balanced steps will be required to improve energy efficiency, by increasing the share of renewables and reducing the level of fossil fuels in the energy mix. Sustainable efforts should be taken to achieve greenhouse gas emission-reduction goals, particularly in energy production, agriculture, transport, industry and the building sector.

The fight against climate change has always been extremely important for Ukraine. Over recent times, especially during 2020, Ukrainians felt the effects of climate change in the shape of large-scale fires in the Chornobyl exclusion zone, floods and droughts in the southern regions. Climate change in Ukraine is leading to serious consequences, including increased risks to human health; a significant reduction in major crop yields; water supply interruptions in the southern and south-eastern regions; land degradation and desertification; the declining viability of forests; destruction of wildlife.

The impact of Russia's invasion of Ukraine is reflected in a staggering account of increasing greenhouse gas emissions. This is due *inter alia* to military activity; toxic spillage and clouds caused by the destruction of industrial and fuel storage facilities; the contamination of water and soil from heavy metals and chemicals from bombs as well as weaponry; risks of radiation release.

S. Chekunova (✉)
Razumkov Centre, Kyiv, Ukraine

© The Author(s), under exclusive license to Springer Nature Switzerland AG 2023   171
M. Kaeding et al. (eds.), *Climate Change and the Future of Europe*, The Future of
Europe, https://doi.org/10.1007/978-3-031-23328-9_39

Nevertheless, Ukraine is committed to fighting climate change and meeting greenhouse gas emission-reduction goals as set out under the Paris Agreement. The country has already made a significant contribution to reducing emissions. More specifically, by 2019 Ukraine had achieved a decrease of 62.4% on the 1990 levels (including land use, land-use change and forestry) and 64.8% on the 1990 levels (excluding land use, land-use change and forestry). This reduction was due largely to structural changes in the economy, brought about by lower production levels in industry and agriculture; decreasing fuel consumption in the energy sector; the global economic crisis in 2008; followed by an adverse impact on the economy in 2014 due to Russia's annexation of Crimea and its occupation of other Ukrainian territory in parts of the Donetsk and Luhansk regions.

At the 2021 UN Climate Conference of the Parties in Glasgow, Ukraine presented its Second Nationally Determined Contribution, aiming to produce a 65% reduction in greenhouse gas emissions by 2030, compared with 1990 (including land use, land-use change and forestry). As part of this conference, Ukraine also joined a number of international initiatives, including a 30% reduction in methane emissions by 2030 compared with 2020 (Global Methane Pledge); ending the use of coal in electricity generation by 2035; large-scale investment in renewable energy (Powering Past Coal Alliance).

Before the war, 70% of electricity generated in Ukraine was low carbon in terms of greenhouse gas emissions. This was achieved through the combination of a high nuclear share (55%), hydro generation (7%) and an increased share of renewable energy (8%). However, Ukraine needs investment and financial support to implement projects that meet the Paris Agreement goals and principles of carbon-free sustainable development. The path to climate neutrality—an economy with net-zero greenhouse gas emissions by 2060—envisages: increasing energy efficiency and the share of renewable energy sources; transitioning to environmentally friendly transport; introducing a circular economy; developing 'smart' networks and communications; expanding the use of bioenergy and synthetic fuels; as well as seeking innovations in the energy sector and hydrogen economy.

Appropriate legislation has been adopted not only to improve energy efficiency in buildings, industry, transport and utilities, but also to reduce and replace the consumption of imported energy resources as much as possible. In Ukraine renewables can play a major role in decarbonisation. According to Ukraine's renewables draft Action Plan, there needs to be an increase in the capacity of renewable generation by 6 gigawatts (from 15.6 gigawatts in 2021 to 21.6 gigawatts in 2030), almost 15 gigawatts of which will be the capacity of solar and wind power plants. The aim is to achieve 25% of electricity generation from renewable sources by 2030, as set out in the National Economic Strategy. With renewables integrating the grids, energy storage is vital to ensure that the power generated from renewable energy sources is accumulated. A law was adopted which defines the use of energy storage systems and regulates the licensing of energy storage activities.

Ukraine has also been identified as a key partner in the European Alliance for Clean Hydrogen for the production and supply (export) of hydrogen given the natural resources, the interconnected infrastructure and the level of technological development. Research is underway in Ukraine to explore the production opportunities of low-carbon price-competitive hydrogen and the supply pathways.

The electricity grids of Ukraine and Moldova have also been successfully synchronised with the European Power System, operated by the European Network of Transmission System Operators for Electricity. This process was expected to take years to accomplish (scheduled for completion in 2023). Yet, Ukraine needs to implement EU legislation to enable market coupling, that is the electricity market operating under EU rules.

However, understandably there are numerous unfavourable factors. As a result of hostilities, the country's infrastructure has been heavily damaged, with the functioning of electricity networks and district heating systems, together with other facilities being far from stable. Ukrainian electricity market segments are operating with heavy distortions (*inter alia* public service obligations, price caps on Day-Ahead Market and Intraday markets, artificial regulation, debts between a number of market players). Under war conditions, it is challenging for potential investors to finance the projects and technologies that would otherwise be contributing to energy transition.

However, it is already time to think about reconstruction and revitalisation of the economy. Ukraine needs to utilise expertise in combining cutting-edge technologies and renewables so as to move towards achieving its carbon-neutral targets. In accordance with the report dealing with assessment of generating capacity adequacy to cover projected demand for electricity by 2031, Ukraine needs at least 2 gigawatts of shunting power (currently only 1200 megawatts are available from thermal and hydro power plants). Since the start of the war, production and consumption of electricity have fallen by 34%, including industrial consumption by 32%. Nuclear power production has fallen by 34%, thermal power production by 28% and electricity generation from renewable energy sources by 72%. The required energy resources' structure will undergo significant changes, primarily due to the economy's advanced electrification (transport, industry, buildings).

Ukrainians admit that climate change is already affecting their lives. The Razumkov Centre's polling of November 2021 shows a level of concern about climate change, with 76.8% of women and 71.2% of men (aged 18–65 and educated) agreeing that climate change is a threat to human life. Despite present hostilities, Ukraine's society and government remain supportive of the fight against climate change, albeit during the current war other priorities must clearly take precedence, such as assisting the Armed Forces, restoring critical infrastructure and dealing with humanitarian issues.

As the EU is fighting to make Europe the first climate-neutral continent in the world by 2050 as it is defined in the European Green Deal, Ukraine has announced its strong commitment to working together with the EU toward green transition. It is envisaged under the 'National Economic Strategy of Ukraine until 2030' that carbon neutrality will be achieved by 2060. In March 2022, the Ukrainian energy system was connected to Europe's continental grid, thus ensuring grid stability and electricity supplies in and out of the country as may be necessary. Although this has been years in the making, the final stages were accelerated, due to the war. In the long run, Ukraine would have the capacity to export surplus electricity to Europe of up to two gigawatts per year. It may well be nuclear energy, or even renewable energy provided certificates of origin are introduced. Before the war broke out, Ukraine had started evaluating the possible impact of the Carbon Border Adjustment Mechanism upon domestic producers based on the timing of a 'carbon tax' being introduced and applied.

Ukraine plans to establish a national Emission Trading System in line with its obligations under the Ukraine-EU Association Agreement. While implementing a greenhouse gas quota trading system in the country, it is necessary to focus on the concept, as provided by EU Directive 87/2003. Ukraine has improved its emissions monitoring system by introducing relevant regulations. The country has submitted proposals to REPowerEU, a plan for the EU to stop using Russian hydrocarbons. This includes energy-saving (including the creation of tax incentives), energy supply diversification and acceleration of the transition to renewables.

Ukraine is able to strengthen the EU's energy security in the areas of electricity supply, use of renewables and diversification of gas imports. Notably, Ukraine will in turn benefit from a new EU Strategy for external energy engagement in a changing world developed by the European Commission. It provides assistance for Ukraine, including the purchase of urgent goods through the Energy Support Fund of Ukraine and support for reforms leading to the future full integration of Ukraine's energy market with the EU. Decarbonisation was one of the country's energy priorities before the war, but Russia's invasion has now made this approach even more urgent. Ukraine intends to have one of the greenest sectors of energy production in Europe after implementation of this decarbonisation strategy.

Energy transition will be based on a strong nuclear sector, an increase in wind and solar energy, together with an increase in hydropower production from 13% of the energy balance to almost 30%. Ukraine's energy strategy also envisages a significant expansion of biomass and biofuel production. According to analysts, the country could generate 6–8 billion cubic metres of biomethane by 2050. Biomethane retains the benefits of natural gas and can be used for heat and electricity generation, in industry and transport. Moreover, it is carbon neutral.

Although Ukraine and its European partners have yet to estimate losses resulting from Russia's invasion, the issue of compensation for damages and payment of relevant war reparations by the aggressor is on the energy market's agenda. The country needs to think about flexible generation sources that can complement the increasing contribution from renewables in the energy mix, within which future sources such as hydrogen and synthetic fuels should be added.

Recommendations

With the EU involving Ukraine in its new initiatives, a clear joint agenda is needed by government authorities, investment banks and all stakeholders, whereby all can work together cohesively and efficiently to succeed in the common decarbonisation goal. Joint efforts with European partners are required to launch a REPower Ukraine initiative to ensure the country's energy supply and recovery after the war.

Ukraine should set up a more ambitious 2030 goal of reducing greenhouse gas emissions by more than 70% when compared with 1990, primarily by: reviewing targets for metallurgy and other industries; as well as increasing the share of renewables in electricity, transportation and heating.

With investment from the EU, Ukraine is set to promote the hydrogen industry, a critical facilitator in energy transition. Hydrogen can potentially diversify the country's fuel mix towards low-carbon options for electricity generation, heavy transportation and some industrial processes.

As a matter of urgency, EU norms and standards must be finalised for incorporation into Ukraine's legislation so as to be fully prepared for EU membership.

The electricity and natural gas markets models should be upgraded to become compatible with EU markets so as to increase trade in electricity and natural gas with EU countries. The pathway toward carbon neutrality cannot be undertaken without overcoming the crisis in energy markets, which slows down the integration of Ukraine's energy systems with those in Europe.

The country's climate goals need to be revised after the war, but they should certainly not be less ambitious, given the growing importance of energy security and energy independence from Russian fossil fuels.

Svitlana Chekunova is a Leading Expert in the Energy Programmes at Razumkov Centre and focuses her research on: international energy markets; energy security; global energy strategies and implications; green transition; climate change; renewables; energy efficiency; and environmental matters.

The Razumkov Centre is a non-governmental think tank founded in 1994, uniting experts in the fields of: economy, energy, law, political sciences, international relations, military security, land relations, sociology, history and philosophy. The Centre is also a member of TEPSA.

The manufacturer's authorised representative in the EU is Springer
Nature Customer Service Centre GmbH, Europaplatz 3, 69115 Heidelberg,
Germany. If you have any concerns regarding our products, please
contact ProductSafety@springernature.com

Printed and bound by CPI Group (UK) Ltd, Croydon, CR0 4YY
24/04/2026
02096360-0001